INTRODUCING RESEARCH METHODOLOGY

Sara Miller McCune founded SAGE Publishing in 1965 to support the dissemination of usable knowledge and educate a global community. SAGE publishes more than 1000 journals and over 800 new books each year, spanning a wide range of subject areas. Our growing selection of library products includes archives, data, case studies and video. SAGE remains majority owned by our founder and after her lifetime will become owned by a charitable trust that secures the company's continued independence.

Los Angeles | London | New Delhi | Singapore | Washington DC | Melbourne

3E

UWE FLICK

INTRODUCING RESEARCH METHODOLOGY

Thinking Your Way Through Your Research Project

Los Angeles | London | New Delhi
Singapore | Washington DC | Melbourne

Los Angeles | London | New Delhi
Singapore | Washington DC | Melbourne

SAGE Publications Ltd
1 Oliver's Yard
55 City Road
London EC1Y 1SP

SAGE Publications Inc.
2455 Teller Road
Thousand Oaks, California 91320

SAGE Publications India Pvt Ltd
B 1/I 1 Mohan Cooperative Industrial Area
Mathura Road
New Delhi 110 044

SAGE Publications Asia-Pacific Pte Ltd
3 Church Street
#10-04 Samsung Hub
Singapore 049483

Editor: Alysha Owen
Assistant editor: Charlotte Bush
Assistant editor, digital: Sunita Patel
Production editor: Rachel Burrows
Copyeditor: Sharon Cawood
Proofreader: Brian McDowell
Indexer: Silvia Benvenuto
Marketing manager: Ben Griffin-Sherwood
Cover design: Shaun Mercier
Typeset by: C&M Digitals (P) Ltd, Chennai, India
Printed in the UK

Library of Congress Control Number: 2019947467

British Library Cataloguing in Publication data

A catalogue record for this book is available from
the British Library

ISBN 978-1-5264-9694-2
ISBN 978-1-5264-9693-5 (pbk)

At SAGE we take sustainability seriously. Most of our products are printed in the UK using responsibly sourced
papers and boards. When we print overseas we ensure sustainable papers are used as measured by the PREPS
grading system. We undertake an annual audit to monitor our sustainability

CONTENTS

EXTENDED CONTENTS

LIST OF FIGURES

LIST OF TABLES

ABOUT THE AUTHOR

Uwe Flick is Professor of Qualitative Research in Social Science and Education at the Freie Universität Berlin, Germany. He is a trained psychologist and sociologist and received his PhD from the Freie Universität Berlin in 1988 and his Habilitation from the Technical University of Berlin in 1994. He has been Professor of Qualitative Research at Alice Salomon University of Applied Sciences in Berlin, Germany and at the University of Vienna, Austria, where he continues to work as Guest Professor. Previously, he was Adjunct Professor at the Memorial University of Newfoundland in St John's, Canada; a Lecturer in research methodology at the Freie Universität Berlin; a Reader and Assistant Professor in qualitative methods and evaluation at the Technical University of Berlin; and Associate Professor and Head of the Department of Medical Sociology at the Hannover Medical School. He has held visiting appointments at the London School of Economics, the École des Hautes Études en Sciences Sociales in Paris, Cambridge University (UK), Memorial University of St John's (Canada), University of Lisbon (Portugal), Institute of Higher Studies in Vienna, in Italy and Sweden, and the School of Psychology at Massey University, Auckland (New Zealand). His main research interests are qualitative methods, social representations in the fields of individual and public health, vulnerability in fields like youth homelessness or migration, and technological change in everyday life. He is the author of *Designing Qualitative Research* (Sage, 2nd edn, 2018) and *Managing Quality in Qualitative Research* (Sage, 2nd edn, 2018), and editor of *The SAGE Handbook of Qualitative Data Analysis* (Sage, 2014), *The SAGE Qualitative Research Kit* (Sage, 2nd edn, 2018), *A Companion to Qualitative Research* (Sage, 2004), *Psychology of the Social* (Cambridge University Press, 1998), *Quality of Life and Health: Concepts, Methods and Applications* (Blackwell Science, 1995) and *La perception quotidienne de la Santé et de la Maladie: Théories subjectives et Représentations sociales* (L'Harmattan, 1993). His most recent publications are the sixth edition of *An Introduction to Qualitative Research* (Sage, 2018), *Doing Grounded Theory* (Sage, 2018), *Doing Triangulation and Mixed Methods* (Sage, 2018) and *The SAGE Handbook of Qualitative Data Collection* (editor, Sage, 2018). In 2019, Uwe Flick received the Lifetime Award in Qualitative Inquiry at the 15th International Congress of Qualitative Inquiry.

PREFACE TO THE THIRD EDITION

Several developments have shaped the context of this book. First, the political and practical relevance of social research has grown. Empirically based knowledge on such issues as the gap between rich and poor, changes in the incidence of diseases and the effects of social disadvantage provide the basis for decision-making, both in policy and in professional practice.

Second, an increasing number of university programs include either introductory or advanced training in the principles and methods of social research. In most cases, this covers questions not only of how to understand existing research, but also of how to conduct research projects (on whatever scale). Sometimes this training is embedded in a course or research-based teaching. Often, however, the research project forms a basis for the final (bachelor, master's, doctoral) thesis and students may be more or less working on their own whilst planning and running their research projects.

Background to the Book

Two background experiences have informed the writing of this book. First, there is my own experience of conducting social research in several fields (including health, youth studies, technological change, ageing and sleep, migration and employment). This experience has taught me a good deal about the problems that arise in research and how to deal with them. Second, there is my experience of teaching social research methods to students and doing social research projects with students in psychology, sociology, social psychology, nursing, public health and education. This experience has taken several forms, including research-based teaching, seminar projects, and supervising numerous bachelor, master's and PhD theses. This work has helped me to discover which examples of other researchers' work most serve to inform what research is about.

Aims of the Book

This book is designed to help readers who are embarking on social research projects. There are, of course, numerous resources on social research already available, including some comprehensive textbooks. In introducing social research, however, comprehensiveness is not necessarily a virtue. Comprehensive treatments tend to be bulky and unwieldy and they can be overwhelming in the detail they present.

In contrast, this book aims to provide the reader with a concise overview. It outlines the most important approaches likely to be used in social research projects. And it provides a good deal of practical information on how to proceed with a project. It also includes guidance on, and reference to, further sources on the subject.

Overview of the Book

The first part of the book will give you an *orientation* to the field of social research. It focuses on issues that come into play as one begins to approach a research project. Chapter 1 provides an introductory overview of what social research is, what you can do with it – and what you can't. It also addresses quantitative and qualitative research as two legs of social research. Chapter 2 gives a brief introduction to the major worldviews behind quantitative and qualitative research. Chapter 3 outlines issues of research ethics in quantitative and qualitative research, including data protection, codes of ethics and the role of ethics committees. The fourth chapter shows how research questions originate and how they can be developed and refined. It considers research questions in the context of both qualitative and quantitative research. This chapter also outlines the role of hypotheses.

The second part of the book deals with *planning and designing* a research project. Chapter 5 shows how to find existing research literature and how to use it in your own project. The sixth chapter provides a short overview of the major steps involved in the research process (for both quantitative and qualitative research). Chapter 7 focuses on the design of quantitative and of qualitative research. First, it provides guidance on how to develop a research proposal and a timescale for your project. In the next step, it discusses key research designs in quantitative and qualitative research. The last part of this chapter deals with sampling – the selection of participants. Some of the major strategies of sampling are described.

In the (new) third part of the book, we turn to the issue of *selecting methods*. The eighth chapter outlines the selection of methods and approaches to be used for pursuing your own research question. A central focus is on the decisions you will need to make at various stages of the research process. Chapter 9 considers ways of combining approaches through triangulation and mixed methods, which are presented as alternatives. The main focus is on triangulation and mixed methods as issues of designing and selecting methods for a specific study.

The fourth part of the book turns to the business of *working with data*. First, the ways of using existing data for one's own study are discussed. Secondary analysis, re-analysis and the use of documents (texts, images, virtual and social media data) are discussed in the new Chapter 10. Methods of collecting new data are the focus of Chapter 11. Surveys, interviews and observation are all discussed here and issues concerning measurement and documentation are outlined. The analysis of quantitative and qualitative data, whether existing or newly created data, is the topic of Chapter 12. This chapter introduces content analysis, descriptive statistics and qualitative analysis, as well as case studies and the development of typologies.

The fifth and final part of the book addresses issues of *reflection and writing* about your project as a whole and presenting its results. Chapter 13 focuses on the evaluation of empirical studies in quantitative and qualitative research. Criteria for evaluation in both areas, as well as questions of generalization, are discussed. This chapter also discusses the limitations of the various methods in quantitative and qualitative research and of each approach in general. The final chapter discusses issues of writing about research. It describes how results in qualitative and quantitative research can be presented and, in particular, how to provide feedback to participants and how to use results in practical contexts and in wider debates.

Features of the Book

Every chapter begins with a *list of objectives*. These specify what I hope you will learn from each chapter. A *navigator* through a research project is also provided before the chapter begins so that you can see at a glance how each chapter fits into the whole. To illustrate, the Chapter 1 Navigator is shown overleaf.

Case studies of *research in the real world* and other material are provided throughout to illustrate methodological issues. At the end of each chapter, you will find checklists for *what you need to ask yourself* about the topics of the chapter and of *what you need to succeed* whilst planning and conducting a research project. These checklists provide readily accessible guidance that can be referred to over and over again as your project progresses. *Key points* and suggestions for *what's next* to *read further* and of *what you have learned* conclude every chapter. A *glossary* explaining the most important terms and concepts used in the text is included at the end of the book.

The navigator

You are here in your project

Orientation

- Why social research?
- Worldviews in social research
- Ethical issues in social research
- From research idea to research question

Planning and design

- Reading and reviewing the literature
- Steps in the research process
- Designing social research

Method selection

- Deciding on your methods
- Triangulation and mixed methods

Working with data

- Using existing data
- Collecting new data
- Analyzing data

Reflection and writing

- What is good research? Evaluating your research project
- Writing up research and using results

The Third Edition – What is New?

The third edition includes a large number of revisions to the existing chapters. They consist of a stronger focus on quantitative research and qualitative research as two legs of social research. Their common features, their individual logics and their differences are more clearly highlighted in the new edition of the book. Issues of research ethics for both legs of social research are given a more prominent place in the logic of the book. The chapter on worldviews in quantitative and qualitative research sets up a stronger epistemological framework for introducing both approaches. The chapter on deciding between methodological approaches in the research process has been straightened up. The core chapters on data collection and data analysis have been extended by including a wider range of methodological approaches. They are complemented by a new chapter on using existing data for your research. The issues about virtual and digital research have been extended by using social media research and Big Data as up-to-date approaches to doing social research. Online and digital research issues have been integrated in the discussion of ethics, design and methods in the other chapters, so that there is no longer an extra chapter on e-research. Finally, the focus on designing students' research projects is now stronger in the book.

I hope that this book and, in particular, its current revised edition will stimulate your curiosity about doing a social research project and, by guiding you through such a project, show you that doing a research project can be an enjoyable and exciting experience.

ONLINE RESOURCES

Get the methods support you need, when you need it: **https://study.sagepub.com/ flickirm3e**.

Organised around the five different parts of the research process, from orientating yourself within social research to writing up and reflecting on research, this book's online resources provide the support you need to think about and understand your entire research project.

For Students

Videos introduce you to the key aspects of the research process, from the theoretical background to presenting your research. Uwe also offers his top tips for improving your research and academic skills.

Case studies showcase methodology in practice, with critical thinking questions to help you develop your ability to interpret and understand research.

Weblinks to guidelines, journal repositories and more help you find other people's research, cite your work correctly and make sure your research meets legal and ethical codes.

Journal articles illustrate how methods can be used in the real world, and the questions accompanying them push you to critically assess research and think about your own practice.

A downloadable **'argument builder'** template walks you step-by-step through how to develop a coherent, robust argument when writing up research or a literature review, helping you improve your communication and critical thinking skills.

For Instructors

Download **PowerPoint slides** featuring figures and tables from the book, which can be customised for use in your own lectures and presentations.

Check your students' understanding with a **test bank** of multiple choice questions related to the key concepts at each stage of the research process that can be used in class, as homework or exams.

Use the free **resource pack** to upload all the lecturer and student resources into your institution's learning management system (e.g. Blackboard or Moodle) and customise the content to suit your teaching needs.

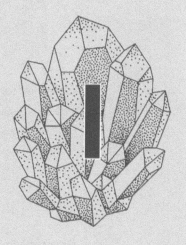

ORIENTATION

Part I of this book has four aims. First, it seeks to introduce social research in general. It considers what social research is, what distinguishes it, what forms it takes, and how it can (and cannot) be used. These are the topics of Chapter 1.

Second, it seeks to lay a foundation for your own research project. In the second chapter, we will address the basic worldviews of (quantitative and qualitative) research and their differences (Chapter 2). In the next step, we then discuss research ethics and what it means to make a research project ethically sound. In the process, we consider codes of ethics and ethics committees, as well as specific problems of online research in this context (Chapter 3). Following that, Chapter 4 introduces the issues of developing research interests and the use of hypotheses. It considers what a research question is and how such questions may be developed. This chapter also considers hypotheses – what they are and how and when they are useful.

These four chapters together also introduce a theme that runs throughout the book, namely the distinction and the relationship between qualitative and quantitative research.

Research Methodology Navigator

You are here in your project

Orientation

- Why social research?
- Worldviews in social research
- Ethical issues in social research
- From research idea to research question

Planning and design

- Reading and reviewing the literature
- Steps in the research process
- Designing social research

Method selection

- Deciding on your methods
- Triangulation and mixed methods

Working with data

- Using existing data
- Collecting new data
- Analyzing data

Reflection and writing

- What is good research? Evaluating your research project
- Writing up research and using results

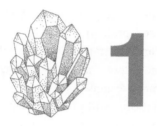

WHY SOCIAL RESEARCH?

How this chapter will help you

You will:

- gain an introductory understanding of social research
- begin to see the similarities and differences between qualitative and quantitative research, and
- appreciate (a) the tasks social research has, (b) what social research can achieve, and (c) what aims you can achieve through it.

What is Social Research?

Increasingly, science and research – their approaches and results – inform public life. They help to provide a basis for political and practical decision-making. This applies across the range of sciences – not only to natural science and medicine, but to social science too. Our first task here is to clarify what is distinctive about social research.

Everyday life and social science

Many of the issues and phenomena with which social research engages also play a role in everyday life. Consider, for example, one issue that is obviously highly relevant to everyday life, namely health. For the most part, health becomes an explicit issue in everyday life only when health-related problems occur or are threatening individuals. Symptoms produce an urge to react and we start to look for solutions, causes and explanations. If necessary, we may go to see a doctor and may end up changing our habits and behaviors – for example, by taking more exercise.

This search for causes and explanations, and people's own experiences, often lead to the development of everyday theories (for example: 'An apple a day keeps the doctor away'). Such theories are not necessarily spelled out explicitly: they often remain implicit. The question of whether everyday explanations and theories are correct or not is usually tested pragmatically: do they contribute to solving problems and reducing symptoms or not? If such knowledge allows the problem at hand to be solved, it has fulfilled its purpose. Then it is not relevant whether such explanations apply to other people or in general. In this context, scientific knowledge (for example, that smoking increases the risk of cancer) is often picked up from the media.

Health, health problems and how people deal with them constitute issues for social research too. But in social science we take a different approach. Analysis of problems is foregrounded and study becomes more systematic. This aims at breaking up routines in order to prevent harmful behaviors – for example, the relation between specific behaviors (like smoking) and specific health problems (such as the likelihood of falling ill with cancer). To achieve such an aim, we need to create a situation free of pressure to act. For example, you will plan a longer period for analyzing the problem, without the pressure of immediately finding a solution for it. Here, knowledge results not from intuition, but from the examination of scientific theories. The development of such theories involves a process of explicitly spelling out and testing relations, which is based on using research methods (like a systematic review of the literature or a survey). For both aims – the developing and testing of theories – the methods of social research are used. The resulting knowledge is abstracted from the concrete example and further developed in the direction of general relations. Unlike in everyday life, here the generalization of knowledge is more important than solving a concrete problem in the particular case. Scientific research is more and more confronted with the expectation that its results have an impact on the field that is studied or on the way a society deals with an issue or (social) problem (see Chandler 2013; Denicolo 2013).

Everyday knowledge and problem solving can of course become the starting points for theory development and empirical research. We may ask, for example, which types of everyday explanations for a specific disease can be identified in interviews with patients.

Table 1.1 presents the differences between everyday knowledge and practices on the one hand, and science and research on the other. It does so on three levels, namely (1) the context of knowledge development, (2) the ways of developing knowledge and the state of the knowledge, which is produced, and (3) the mutual relations between everyday knowledge and science.

Table 1.1 Everyday knowledge and science

	Everyday knowledge and practices	Science and research
Context of knowledge (production)	Pressure to act Solving of problems is the priority: • routines are not put to question • reflection in case of practical problems	Relief from a pressure to act Analyzing of problems is the priority: • systematic analysis • routines are put to question and broken down
Ways of knowledge (production)	Intuition Implicit development of theories Experience-driven development of theories Pragmatic testing of theories Check of solutions for problems	Use of scientific theories Explicit development of theories Methods-driven development of theories Methods-based testing of theories Use of research methods
State of knowledge	Concrete, referring to the particular situations	Abstract and generalizing
Role of knowledge	Understanding and maybe problem-solving in concrete contexts and situations	Impact on social or societal problems and their solution
Relation of everyday knowledge and science	Everyday knowledge can be used as a starting point for theory development and empirical research	Everyday knowledge is increasingly influenced by scientific theories and results of research

What, then, characterizes social research in dealing with such issues? Here we may itemize a number of characteristics, each of which is explored further in this book. They are:

- Social research approaches issues in a systematic and, above all, empirical way.
- For this purpose, you will develop research questions (see Chapter 4).
- For answering these questions, you will use, collect and analyze data.
- You will use, collect and analyze these data by applying research methods (see Chapters 10, 11 and 12).

- The results are intended to be generalized beyond the examples (cases, samples, etc.) that were studied (see Chapter 13).
- From the systematic use of research methods and their results, you will derive descriptions or explanations of the phenomena you study.
- For a systematic approach, time, freedom and (other) resources are necessary (see Chapter 7).

Definition of social research

As we shall see, there are different ways of doing social research. First, though, we can develop a preliminary general definition of social research derived from our discussion so far.

Social research is the systematic analysis of research questions by using empirical methods (e.g. of asking, observing, re-using and analyzing data). Its aim is to make empirically grounded statements that can be generalized or to test such statements. Various approaches can be distinguished as can a number of fields of application (health, education, poverty, etc.). Various aims can be pursued, ranging from an exact description of a phenomenon, to its explanation, or to the evaluation of an intervention or institution.

This definition provides an orientation for our journey through social research in this book.

The Tasks of Social Research

We can distinguish three main tasks for social research. To do so, we use the criterion of how the results of social research may be used.

Knowledge: description, understanding and explanation of phenomena

A central task of social research originates from scientific interests, which means that the production of knowledge is prioritized. Once a new phenomenon, such as a new disease, arises, a detailed description of its features (symptoms, progression, frequency, etc.) on the basis of data and their analysis becomes necessary. The first step can be a detailed description of the circumstances under which it occurs or an analysis of the subjective experiences of the patients. This will help us to understand the contexts, effects and meanings of the disease. Later, we can look for concrete explanations and test which factors trigger the symptoms or the disease, which circumstances or medications have specific influences on its course, etc. For these three steps – (a) description, (b) understanding and (c) explanation – the scientific interest in new knowledge is dominant. Such research contributes to basic research in that area. Here, science and scientists remain the target group for the research and its results.

Practice-oriented research: applied and participative research

Increasingly, social research is being conducted in practical contexts such as hospitals and schools. Here, research questions focus on practices – those of teachers, nurses or physicians – in institutions. Or they focus on the specific conditions of work in these institutions – routines in the hospital or teacher–student relations, for example. The results of applied research of this kind are also produced according to rules of scientific analysis. However, they should become relevant for the practice field and for the solution of problems in practice.

A special case here is participatory action research. Here, the changes initiated by the researcher in the field of study do not come only after the end of the study and the communication of its results. The intention is rather to initiate change *during* the process of research and by the very fact that the study is being done. Take, for example, a study of nursing with migrants. A participatory action research study would not set out merely to describe the everyday routines of nursing with migrants. Rather, it would initiate the process of research immediately in those everyday routines. It would then feed back to participants the information gathered in the research process.

This changes the relationship between researcher and participant. A relation which is usually monologic in traditional research (e.g. the interviewees unfold their views, the researchers listen) becomes dialogic (the interviewees unfold their views, the researchers listen and make suggestions for how to change the situation). A subject–object relation turns into a relation between two subjects – the researcher and the participant. The evaluation of the research and its results is no longer focused solely on the usual scientific criteria (as will be discussed in Chapter 13). Rather, the question of the usefulness of the research and its results for the participant becomes a main criterion. Research is no longer just a knowledge process for the researchers, but rather a process of knowledge, learning and change on both sides.

Grounding political and practical decisions

Since the middle of the twentieth century, social research has become more important as a basis for decisions in practical and political contexts. In most countries, regular surveys in various areas are common practice; reports on health, on poverty and on the situation of the elderly and of youth and children are produced, often commissioned by government. In many cases, such monitoring does not involve extra research, but rather summarizes existing research and results in the field. But, as the PISA studies or the HBSC study (Hurrelmann et al. 2003) show, in areas like health, education and youth, additional studies do sometimes contribute to the basis of these reports. In the HBSC study, representative data about 11- to 15-year-old adolescents in the population are collected. At the same time, case studies with purposefully selected cases are included. Where data from representative studies are not available or cannot be expected, sometimes only case studies provide the basis for data.

In many areas, decisions about establishing, prolonging or continuing services, programs or institutions are based on evaluations of existing examples or experimental programs. Here, social research not only provides data and results as a basis for decisions, but also makes assessments and evaluations – by, for example, examining whether one type of school is more successful in reaching its goals than a different type. Therefore, the potential for implementation of research results, and more generally the impact of research beyond academia, become more important. Chandler (2013, p. 3) states:

> The context within which that impact takes place is broad beyond academia in the realms of society, economy, public policy or services, health, the environment or quality of life. The outcomes or indicators of impact encompass the individual, community or global levels and are the application of new knowledge or understanding in the development of policy, creation of products or services.

Table 1.2 summarizes the tasks and research areas of social research outlined above, using the context of health as an example.

What Can You Achieve with Social Research?

In the areas just mentioned, we can use social research to:

- explore issues, fields and phenomena and provide first descriptions
- discover new relations by collecting and analyzing data
- provide empirical data and analyses as a basis for developing theories

Table 1.2 Tasks and research areas of social research

Research area	Features	Aims	Example	Studies refer to
Basic research	Development or testing of theories	General statements without a specific link to practices	Trust in social relationships	Random sample of students or unspecific groups
Applied research	Development or testing of theories in practical fields	Statements referring to the particular field Implementation of results	Trust in doctor–patient relations	Doctors and patients in a specific field
Participatory action research	Analyzing fields and changing them at the same time	Intervention in the field under study	Analysis and improvement of nursing for migrants	Patients with a specific ethnic background, for example, who are (not sufficiently) supported by existing home care services
Evaluation	Collection and analysis of data as a basis for assessing the success and failure of an intervention	Assessment of institutional services and changes	Improvement of the trust relations between doctors and patients in a specific field with better information	Patients in a specific field
Health monitoring	Documentation of health-related data	Stocktaking of developments and changes in the health status of the population	Frequencies of occupational diseases	Routine data of health insurance

- test existing theories and stocks of knowledge empirically
- document the effects of interventions, treatments, programs, etc. in an empirically based way
- provide knowledge (i.e. data, analyses and results) as an empirically grounded basis for political, administrative and practical decision-making.

What is social research unable to do and what can you do with it?

Social research has its limits. For example, the aim of developing a single grand theory to explain society and the phenomena within it, which also withstands empirical testing, could not be achieved. And there is no one method for studying all relevant phenomena. Moreover, social research cannot be relied upon to provide immediate solutions for current, urgent problems. On all three levels, we have to rein in our expectations of social research and pursue more realistic aims.

What we can aim to do is develop, and even test empirically, a number of theories. They can be used to explain certain social phenomena. We can also continue to develop a range of social science methods. Researchers can then select the appropriate methods and apply them to the problems they wish to study. Finally, social research provides knowledge about details and relations, which can be employed to develop solutions for societal problems.

Quantitative and Qualitative Research

We need now to turn to the distinction between qualitative and quantitative research. This distinction is only briefly introduced here, but will feature frequently and will be spelled out in more detail throughout this book. The notions of 'qualitative research' and 'quantitative research' are umbrella terms for a number of approaches, methods and theoretical backgrounds on each side. That is, each of these two terms in fact covers a wide range of procedures, methods and approaches. Nevertheless, they are useful. Here, therefore, we develop an outline of the two approaches (for more details, see Bryman 2016 and Flick 2018a) and consider what characterizes each.

Quantitative research

Quantitative research can be characterized as follows. In studying a phenomenon (e.g. the stress of students), you will start from a concept (e.g. a concept of stress), which you spell out theoretically beforehand (e.g. in a model of stress, which you set up or take from the literature). For the empirical study, you will formulate a hypothesis (or several hypotheses), which you will test (e.g. that for students in humanities, university is more stressful than for students in the natural sciences). In the empirical project, the procedure of measurement has high relevance for finding out differences among persons concerning the characteristics you study (e.g. there are students with more and less stress).

In most cases, we cannot expose a theoretical concept immediately to measurement. Rather, we have to find indicators that permit a measurement in place of the concept. We may say that the concept has to be *operationalized* in these indicators. In our example, you could operationalize stress before an exam by using physiological indicators (e.g. higher blood pressure) and then apply blood pressure measurements. More often, researchers operationalize research through using specific questions (e.g. 'Before exams, I often feel under pressure') with specific alternatives of answering (as in the example in Figure 1.1).

Data collection is designed in a standardized way (e.g. all participants in a study may be interviewed under the same circumstances and in the same way). The methodological ideal is the kind of scientific measurement achieved in the natural sciences. By standardization of the data collection and of the research situation, the criteria of reliability, validity and objectivity (see Chapter 13) can be met.

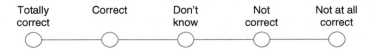

Figure 1.1   Alternatives for answering on a Likert scale

Quantitative research is interested in causalities – for example, in showing that stress before an exam is caused by the exam and not by other circumstances. Therefore, you will create a situation for your research in which the influences of other circumstances can be excluded as far as possible. For this purpose, instruments are tested for the consistency of their measurement, such as in repeated applications. The aim of the study is to achieve generalizable results: that is, your results should be valid beyond the situation in which they were measured (the students also feel the stress or have the higher blood pressure before exams when they are not studied for research purposes). The results from the group of students that participated need to be transferable to students in general. Therefore, you will draw a sample, which you select according to criteria of representativeness – the ideal case is a random sample (see Chapter 7 for this) – from the population of all students. This will mean that you can generalize from the sample to the population. Thus, the particular participants are relevant not as individuals (how does the student Joe Bauer experience stress before exams?) but rather as typical examples. It is not so much the students' entire situation, but rather their specific (e.g. physiological) reactions to a certain condition (a coming exam) that are relevant.

The emphasis on measurement, as in the natural sciences, relates to an important research aim, namely replicability – i.e. the measurement has principally to be able to be repeated, and then, provided the object under examination has not itself changed, to produce the same results. In our example: if you measure blood pressure for the same student before the exam repeatedly, the measured values must be the same – except if there are good reasons for a difference, such as if blood pressure rises as the exam gets closer.

Quantitative research works with numbers. To return to our example: because measurement produces a specific figure for blood pressure, the alternatives for answering in Figure 1.1 can be transformed into numbers from 1 to 5. These numbers make a statistical analysis of the data possible (see Bryman 2016 for a more detailed presentation of these features of quantitative research). Kromrey (2006, p. 34) defines the 'strategy of the so-called quantitative research' as 'a strictly goal-oriented procedure, which aims for the "objectivity" of its results by a standardization of all steps as far as possible and which postulates intersubjective verifiability as the central norm for quality assurance'.

The participants may experience the research situation as follows. They are relevant as members of a specific group from which they were selected randomly. They are confronted with a number of predefined questions, for which they also have a number of predefined answers, of which they are expected to choose only one. Information beyond these answers, as well as their own assumptions, subjective states or queries and comments on the questions or the issue, are not part of the research situation.

Qualitative research

Qualitative research sets itself other priorities. Here, you normally do not necessarily start from a theoretical model of the issue you study and refrain from hypotheses and operationalization. Also, qualitative research is not modeled on measurement as found in the natural sciences. Finally, you will be interested neither in standardizing the research situation as far as possible nor in guaranteeing representativeness by the random sampling of participants.

Instead, qualitative researchers select participants purposively and integrate small numbers of cases according to their relevance. Data collection is designed much more openly and aims at a comprehensive picture made possible by reconstructing the case under study. Thus, fewer questions and answers are defined in advance; there is greater use of open questions. The participants are expected to answer these questions spontaneously and in their own words. Often, researchers work with narratives of personal life histories.

The example in Box 1.1 demonstrates how a research question for which not much empirical research was available can be addressed in a student project with qualitative methods.

Qualitative research addresses issues by using one of the following three approaches. It aims (a) at grasping the subjective meaning of issues from the perspectives of participants (e.g. what does it mean for interviewees to experience their university studies as a burden?). Often, (b) latent meanings of a situation are in focus (e.g. which are the unconscious aspects or the underlying conflicts that influence the experience of stress for the student?). It is less relevant to study a cause and its effect than to describe or reconstruct the complexity of situations. In many cases, (c) social practices and the life world of participants are described. The aim is less to test what is known (e.g. an existing theory

---

Box 1.1   Research in the real world

A student's research study on the emancipation of refugee women in Germany

For her empirical bachelor thesis in education, Jackleen Khazal (2018) was interested in exploring the influence of experiences of being a refugee on Syrian women's emancipation in Germany. To answer her research question, she interviewed four women of 20, 28, 36 and 48 years who had come to Germany on differing routes from Syria. Khazal used the episodic interview (see Chapter 11) with an interview guide comprising five topics – the interviewee and her experience as refugee; everyday life in Syria and Germany; her understanding of emancipation; life in Syria and Germany and societal demands; and ideas about her future in and perspective on Germany. The data were analyzed using thematic coding (see Chapter 12). The results show specific understandings of the interviewed women and differences among them.

---

or hypothesis) than to discover new aspects in the situation under study and to develop hypotheses or a theory from these discoveries. Therefore, the research situation is not standardized; rather it is designed to be as open as possible. A few cases are studied, but these are analyzed extensively in their complexity. Generalization is an aim not so much on a statistical level (generalization to the level of the population, for example) as on a theoretical level. (For a more detailed presentation of these features, see Flick 2018.)

The participants in a study may experience the research situation as follows. They are involved in the study as individuals, who are expected to contribute their experiences and views from their particular life situations. There is scope for what they see as essential, for approaching questions differently and for providing different kinds of answers with different levels of detail. The research situation is designed more as a dialogue, in which probing, new aspects, and their own estimations find their place.

Figure 1.2 displays the various levels on which social research can approach a phenomenon under study. If we take failure in school exams as a phenomenon, we can use quantitative research and study distributions and correlations of this phenomenon in big samples, e.g. asking for differences in failure rates between small town and big city schools. We can use qualitative methods to develop a theory or a typology based on smaller samples, asking participants for their experience with failing in exams and subjective views about it. And, in a more radical way, we can do an empirical case study asking for how one student experiences and explains failing in school exams, what has led to this result and how the individual's life continues after it. This approach allows a detailed description and in-depth understanding of a case of failing, which then can be complemented by other case studies.

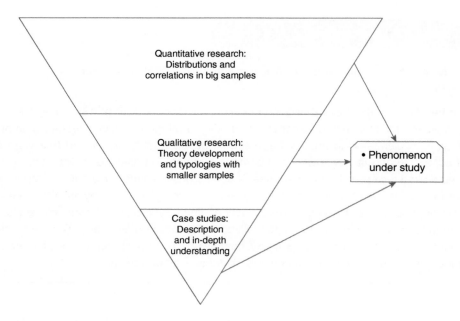

Figure 1.2   Levels of approaching a phenomenon under study

Differences between quantitative and qualitative research

From the above outlines of features of both approaches, some of the main differences in assessing what is under study (issue, field and persons) have become evident. These are summarized in Table 1.3.

Table 1.3   Differences between quantitative and qualitative research

	Quantitative research	Qualitative research
Theory	As a starting point to be tested	As an end point to be developed
Case selection	Oriented on (statistical) representativeness, ideally random sampling	Purposive according to the theoretical fruitfulness of the case
Data collection	Standardized	Open
Analysis of data	Statistical	Interpretative
Generalization	In a statistical sense to the population	In a theoretical sense

Common aspects of quantitative and qualitative research

Despite the differences, the two approaches have some points in common. In both approaches, you:

- use, produce, and analyze data (see Chapters 10 to 12)
- work systematically by using empirical methods (see Chapters 11 and 12)
- aim at generalizing your findings – to situations other than the research situation and to persons other than participants in the study (see Chapter 13)
- pursue certain research questions, for which the selected methods should be appropriate (see Chapter 4)
- should answer these questions using a planned and systematic procedure (see Chapter 7)
- have to check your process of research for ethical acceptability and appropriateness (see Chapter 3)
- have to make your process of research transparent (i.e. understandable for the reader) in presenting the results and the ways that lead to them (see Chapter 14).

Advantages and disadvantages

An advantage of quantitative research is that it allows for the study of a large number of cases for certain aspects in a relatively short time frame and its results have a high degree of generalizability. The disadvantage is that the aspects that are studied are not necessarily the relevant aspects for participants and the context of the meanings linked to what is studied cannot be sufficiently taken into account.

An advantage of qualitative research is that detailed and exact analyses of a few cases can be produced, in which participants have much more freedom to determine what is relevant for them and to present it in its contexts. The disadvantage is that these analyses often require a lot of time and you can generalize results to the broad masses in only a very limited way.

Synergies and combinations

The strengths and weaknesses just mentioned provide a basis for deciding which methodological alternative you should select for your specific research question (see Chapter 8). At the same time, we should remember that it is possible to combine qualitative and quantitative research (as is explored in more detail in Chapter 9) with the aim of compensating for the limitations and weaknesses of each approach and producing synergies between them.

Doing Research On Site, Doing it Online or Combining The Two

In the last two decades or so, a new trend has arisen which has considerably extended the reach of social research. With the development of the Internet, both qualitative and quantitative approaches can now be used in new contexts.

Traditionally, interviews, surveys and observations (see Chapter 11) have mostly been done on site. You make appointments with your participants, meet them at a specific time and location, and interact with them face to face or send them your questionnaire by mail and they return it in the same way. This kind of research has its limitations. Sometimes, practical reasons will make these encounters difficult: participants live far away, are not ready to meet researchers, or are relevant for your study as members of a virtual community.

These limitations can sometimes be overcome if you decide to do your study online. Quantitative and qualitative methods have been adapted to online research. E-mail or online interviews, online surveys and virtual ethnography are now part of the methodological toolkit of social researchers. This means not so much (or, at least, not only) that you apply social science methods to study (the use of) the Internet, but rather that you use the Internet to apply your methods for answering your research questions. In particular, the new forms of communication in the context of social media provide new options for doing social research, communicating in and about. Even if studies are mainly conducted on site, the Internet can be a helpful tool and resource. In many cases, traditional forms of data such as interviews or surveys and digital data are combined (Fielding 2018). They also facilitate doing research collaboratively (see Chapters 8–12 for details).

Why and How Research Can Be Fun

For many students, completing courses in research methods and statistics seems to be nothing more than an unpleasant duty; it seems you have to go through this, even if you do not know why and for what purpose. To learn methods can be exhausting and painful. If the whole enterprise leads to a difficult written test at the end, sometimes any excitement is submerged by the stress of the exam. To apply methods can be time-consuming and challenging.

However, the systematic nature of the procedures and the concrete access to practical issues in empirical research in your studies and later in professional work (as a sociologist, social worker and the like) may provide new information. You may discover new insights in the analysis of your data. Interviews, life histories or participant observation can reveal much about concrete life situations or about how institutions function. Sometimes these insights come as surprises, which may give you the chance to overcome your prejudices and limited perspectives on how people live and work. And you will learn a lot about how life histories develop or about what happens in practical work in institutions or in the field.

In most research processes, you will learn a lot not only about the participants, but also about yourself – especially if you work with issues such as health, stress at university, the impact of social discrimination, etc. in concrete life situations. In particular, in the context of theoretically ambitious studies and their contents, working with empirical data can form not only an instructive alternative or complement to theory, but also a concrete link between theory and everyday real-life problems.

Working with other people can be an enriching experience, and if you have the chance to do your research among a group of people – a research team or a group of students – that will be a good way out of the isolation students sometimes experience. For many students, work with technical devices, computers, programs and data can be satisfying and a lot of fun. For example, using the communication forms in social media for research purposes will provide new experiences of social networking and being up to date in the context of using new media in a professional way. And in the end you will have concrete products at hand: examples, results, what they have in common and how they are different for a variety of people, and so on.

Finally, to work on an empirical project requires working on one issue in a sustained way. This is good practice, given that many students' experiences today are characterized more by 'bits and pieces' work. Empirical research in your fields of study can also be a test of how much you like those fields. If the test ends positively, it can reassure you in your decision to become, for example, a social worker or a psychologist.

All in all, research experiences (such as selecting, accessing and interviewing or observing people) can provide experiences and insights that may help you in other, more practice-oriented fields of work after finishing your studies. For highlighting this transferrable benefit of knowing about research, the research examples in the text are introduced as 'Research in the real world' throughout the chapters.

What You Need to Ask Yourself

Knowledge about social research helps in two ways. It can provide the starting point and basis for doing your own empirical study, such as in the context of a thesis or of later professional work in sociology, education, social work, etc. And it is also necessary for understanding and assessing existing research and perhaps for being able to build an argument on such research. For both, we can formulate a number of guideline questions, which allow a basic assessment of research (in the planning of your own or reading other researchers' studies). These are shown in Box 1.2.

Box 1.2   What you need to ask yourself

Understanding social research

1   What is studied exactly?
   - What is the issue and what is the research question of the study?

2   How is it ensured that the research really investigates what is supposed to be studied?
   - How is the study planned, which design is applied or constructed, and how are biases prevented?

3   What is represented in what is studied?
   - Which claims of generalization are made and how are they fulfilled?

4   Is the execution of the study ethically sound and theoretically grounded?
   - How are participants protected from any misuse of the data referring to them?
   - What is the theoretical perspective of the study?

5   Which methodological claims are made and fulfilled?
   - Which criteria are applied?

6   Does the presentation of results and of the ways they were produced make transparent for the reader how the results came about and how the researchers proceeded?
   - Is the study transparent and consistent in its presentation?

7   Is the chosen procedure convincing?
   - Are the design and methods appropriate for the issue under study?

8   Does the study achieve the degree of generalization that was expected?

These guideline questions can be asked regardless of the specific methodology that has been chosen and can be applied to the various methodological alternatives. They are relevant for both qualitative and quantitative studies and can be used for assessing a case study, as well as for a representative survey of the population of a country. They offer a framework for observations as well as for interviewing or for the use of existing data and documents (see Chapters 10 and 11 for more detail on this).

What You Need to Succeed

This chapter should have given you a rough orientation in the field of social research. This should allow you to contextualize what the subsequent chapters will treat in more detail. This orientation should be a first step in helping you later to take on your own research and study. The following questions (see Box 1.3) should help you to check what to take home of this chapter for the steps you will take next.

Box 1.3 What you need to succeed

Understanding social research

1   Do I understand what social research is about?
2   Am I aware of the differences between scientific and everyday knowledge?
3   Do I see what can be reached with social research and what cannot?
4   Do I understand the differences between basic and applied research?
5   Do I have a rough idea about the differences between qualitative and quantitative research?
6   Do I see the common aspects of both forms of research?
7   Do I have an idea when research should be done online and that this follows similar principles as other forms of social research?

What you have learned

- Social research is more systematic in its approach than everyday knowledge.
- Social research can have various tasks: research may be focused on knowledge, practice and consulting.
- Quantitative research and qualitative research offer different approaches. Each has strengths and limitations in what can be studied.
- Quantitative and qualitative research can mutually complement each other.
- Both quantitative and qualitative research can be applied on site and online.
- We can identify common features across the various approaches.

What's next

The first and fifth texts listed below provide further detail on quantitative research and include some chapters on qualitative methods, too; the second, third and fourth books give more insight into the variety of qualitative research methods:

Bryman, A. (2016) *Social Research Methods*, 5th edn. Oxford: Oxford University Press.

Flick, U. (2014) *The SAGE Handbook of Qualitative Data Analysis*. London: Sage.

Flick, U. (2018) *An Introduction to Qualitative Research*, 6th edn. London: Sage.

Flick, U. (2018) *The SAGE Handbook of Qualitative Data Collection*. London: Sage.

Neuman, W.L. (2014) *Social Research Methods: Qualitative and Quantitative Approaches*, 7th edn. Essex: Pearson.

The following case study can be found in the online resources. It should give you an idea of how a relevant problem is selected and turned into a research project. It is only meant to show what relevant research means and is a first orientation for what we address in more detail in the later chapters of the book:

Winters, K. and Carvalho, E. (2014) 'The Qualitative Election Study of Britain: Qualitative Research Using Focus Groups', *SAGE Research Methods Cases*. doi: 10.4135/9781446273050135099945.

The following article can be found in the online resources, too. It should give you an idea of how a relevant problem may be identified as a starting point for a research project:

Kelly, K. and Caputo, T. (2007) 'Health and Street/Homeless Youth', *Journal of Health Psychology*, 12(5): 726–36.

Research Methodology Navigator

You are here in your project

Orientation

- Why social research?
- Worldviews in social research
- Ethical issues in social research
- From research idea to research question

Planning and design

- Reading and reviewing the literature
- Steps in the research process
- Designing social research

Method selection

- Deciding on your methods
- Triangulation and mixed methods

Working with data

- Using existing data
- Collecting new data
- Analyzing data

Reflection and writing

- What is good research? Evaluating your research project
- Writing up research and using results

WORLDVIEWS IN SOCIAL RESEARCH

─How this chapter will help you─

You will:

- have an idea about the differing worldviews behind doing social research
- have a first impression of the different epistemological backgrounds of social research
- see which backgrounds are taken as a starting point mainly by qualitative research and which ones are more relevant for quantitative research
- see the advantages and limits of epistemological backgrounds, and
- be able to use these for evaluating existing research and for planning your own study.

Towards the end of Chapter 1, I showed some of the differences between qualitative and quantitative research. These differences do not only refer to the practical and technical issues of planning a study (see Chapter 6), but reveal different understandings of knowledge and of the relation between social research and reality. This chapter will very briefly outline some of the epistemological backgrounds that social research is based on when studying an issue and which are referred to in research practice. These backgrounds have an impact on the way issues are identified for research, on how they are turned into problems to be studied, and on how they are studied. They also have a major influence on what is seen as an appropriate method for doing social research and on how methods should be applied in a study. As they are sometimes seen as strictly opposite understandings of research, they can also be discussed as worldviews in social research.

Positivism

Positivism as an epistemological program goes back to Auguste Comte. He emphasized that the sciences should avoid speculative and metaphysical approaches and instead concentrate on studying the observable facts ('positiva'). Emile Durkheim emphasized that social facts exist outside individual awareness and rather in collective representations. Positivism as an epistemological program is characteristic of the natural sciences. Bryman (2016, p. 24) summarizes several assumptions of positivism:

1   Only phenomena confirmed by the senses can be warranted as knowledge (phenomenalism).
2   Theories are used to generate hypotheses that can be tested and allow explanations of laws to be assessed (deductivism).
3   Knowledge can be produced by collecting facts that provide the basis for laws (inductivism).
4   Science must and can be conducted in a way that is value-free and thus objective.
5   There is a clear distinction between scientific and normative statements.

This means that phenomena for social research have to be accessible by the senses. In research, theories are the starting points from which hypotheses are deduced and which shall be tested. The research starts from facts that can be collected and from which theories can be inductively derived. The research is oriented on the ideal of objectivity, which means it should be independent of the person of the individual researcher and free of values. Finally, research aims at making scientific statements (it could be shown that X leads to Y) and refrains from normative and evaluative statements.

   Positivism is often associated with realism. Both assume that the natural and social sciences should and can apply the same principles to collecting and analyzing data. They also postulate that there is a world out there (an external reality) separate from our (or the researchers') descriptions of it. This means that we should try to collect facts in and about the world and that this is possible. A collection of attitudes towards a political party may be based on the assumption that the attitude can be seen as a fact and thus be measured (see Chapter 11). The consequence of such a position is that social research is often committed to ideals of measurement and objectivity, rather than reconstruction and interpretation. Thus, the research situation is conceived as independent of the individual researchers who collect and analyze the data, so that the 'facts' (to use a positivist term) above can be collected in an objective way. Therefore, research situations are standardized as far as possible for this purpose. On this view, research should strive for *representativeness* in what is studied, therefore random sampling is often the ideal. Use of the word 'positivism' is often criticized: as Hammersley and Atkinson (1995, p. 2) note, 'all one can reasonably infer from unexplicated usage of the word "positivism" in the social research literature is that the writer disapproves of whatever he or she is referring to'. Qualitative researchers, for example, often use the concept of 'positivism' as something they distinguish their own research from. At the same time, this concept is only seldom spelt out in social science discussions. Positivism is a worldview behind doing quantitative research driven by the idea of transferring the standards of the natural sciences to the social sciences.

Critical Rationalism

Positivism has been much disputed in the development of the social sciences. A more current worldview is the approach of critical rationalism, which goes back to Karl Popper (1971).

Basic assumptions

Critical rationalism is much more skeptical of the idea that knowledge can be derived from observation. It is also skeptical of the testing of hypotheses and theories and, finally, of the aim of gaining secure knowledge: 'Rationalism holds the point that we are able with our rationality to perceive the reality and to act appropriately. Classical rationalism assumes that it is possible to arrive at secure knowledge. Critical rationalism denies this possibility' (Albert 1992, p. 177).

According to Popper (1971, p. 88), the starting point for scientific work is: 'Knowledge does not start from perception or observations or the collection of data or facts, but it starts, rather, from *problems*' (emphasis in original). Thus, scientific studies do not begin with theories or methods but with problems, for which solutions have to be found or for which our knowledge is not sufficient. In this context, the distinction between the context of discovery (How is a hypothesis or theory for explaining a problem developed?) and the context of justification (Is the hypothesis or theory correct or not?) becomes relevant. According to Popper, a central question for science and epistemology is whether and when you can conclude from an observation that an explanation (hypothesis or theory) is correct or not. Popper is not interested in the context of *discovery* of scientific hypotheses but only in the context of *justification* of hypotheses and laws. The aim is not *verification* of scientific statements by empirical observations but to search observations, which put them into question (in the sense of a *falsification*): All statements have to be checked against experience and have to *corroborate* in confrontation with reality. All statements of an empirical science – if they are inappropriate – must be able to fail against experience in principle. The postulates of Popper's deductive-empirical model in this context are:

1   Scientists should construct hypotheses as daring as possible, which have a high level of information and should
2   make them undergo acid tests of corroboration, i.e. a number of empirical tests in a variety of situations.
3   Those meaningful hypotheses, which have withstood the attempts of falsification are kept (tentatively); falsified hypotheses are eliminated (Diekmann 2007, p. 175).

This means that scientific studies are not used for confirming but for falsifying statements by empirical testing of hypotheses. Knowledge is seen as tentative: that is a theory has been corroborated – if the hypotheses derived from it could not be falsified – until a contradictory example is found. Scientific studies, according to this worldview, do not pursue the aim of discovering (new) phenomena or of developing (new) theories. Scientific methods and studies are employed for critically examining existing theories by testing hypotheses derived from them. Popper's aim is to develop a normative methodology, according to which a testing of hypotheses in a scientific way has to be done. The program of critical rationalism is one of the major worldviews behind quantitative research, which not only determines the use of hypotheses in it but also the criteria for judging the quality of quantitative research (see Chapter 13).

Concurrent research programs

But what if different ways (research approaches, methods) of testing hypotheses come about or if various hypotheses or theories exist for explaining the same problem? What if many studies with one approach have corroborated the hypothesis, but one study with a different approach has falsified it? To answer this question without putting the whole approach of critical rationalism into question, other representatives (e.g. Lakatos 1980) assume the existence of concurrent research programs, which may follow different approaches. It should then be discussed how research programs can come to stay or prevail against each other.

Paradigm shifts

To address this problem, Kuhn (1970) formulated his thesis of the shift in scientific 'paradigms' (research programs). Here, he calls into question the idea that scientific progress mainly results from the empirical falsification of one specific or several hypotheses. Rather, progress happens when new views prevail in the sense of scientific revolutions (e.g. the Copernican shift or that Einstein's relativity theory has replaced the classical mechanism). 'A paradigm formulates the maxims and principles on which the scientific practice in a discipline is oriented. Among other things, a paradigm defines the framework of the accepted research methods' (Meuser 2003, pp. 92–3).

However, this also puts the neglect of the context of discovery in Popper's approach into question, as these shifts have been inaugurated in this area and not in the context of justification due to an empirical testing and falsification of hypotheses. Although the notion of paradigm was basically developed for examples coming from the natural sciences, paradigms and paradigm shifts are mentioned quite often in the social sciences too.

Critical rationalism is the major epistemological fundament for quantitative research, which exposes assumptions (hypotheses) to the acid tests in a number of empirical tests, as mentioned above. Thus, a program of skepticism about the concepts of the sciences has been formulated, which have to be put to the test empirically. In developments that could be discussed here only briefly, critical rationalism has become the most important worldview in quantitative research in the social sciences.

Interpretative Paradigm

In the context of rediscovering qualitative methods in the 1960s and 1970s, the basic assumptions of this worldview have been fundamentally put into question. A major argument was the discussion about normative and interpretative paradigms, which was started by Wilson (1970).

Tacit assumptions of critical rationalism – normative paradigm

In this discussion, the understanding of research in critical rationalism has also been labeled a 'normative paradigm'. It 'tacitly' assumes that 'the researcher's knowledge is sufficient for formulating relevant hypotheses', which again 'is based on two – mostly tacit – assumptions:

1   The researchers' and the participants' everyday knowledge relevant to the research question are identical.
2   Participants follow the social rules that are relevant in the area under study in a consistent and unambiguous way; there is no scope for interpretation in the norms: rules of ascribing meaning, for example, are clear for all who live in the social world – all actors know them and are able to apply them in all situations in the same way.' (Kelle 1994, p. 49)

It is assumed that the researchers 'know' the participants' life world sufficiently to formulate a hypothesis for testing their theoretical assumptions in a way that covers the aspects in the real world which are relevant to the test. Further assumptions are that, for example, in a question in a questionnaire (which is used for testing this hypothesis) the understandings on the part of researchers and participants are identical (i.e. both mean and understand the same by it). Thus, it is assumed that in the participants' everyday life, meaning linked with a concept or formulation is not completely different to the researchers' view. And finally, it is assumed that all participants who have responded to a question (e.g. 'Before exams, I often feel under pressure') by ticking 'totally correct' (see Figure 1.1 in Chapter 1) mean the same by this response as the researchers, and that you can see and summarize their responses as identical.

Interpretative paradigm – basic assumptions

As an alternative to this normative paradigm in the social sciences, Wilson (1970) formulated the 'interpretative paradigm', which puts the tacit assumption just mentioned into question. This 'paradigm' includes a variety of theories and approaches of qualitative research. Here, it is not assumed that rules and meanings are clear for all participants in the same way and that researchers' understanding of an issue can be assumed to be that of the participants without further ado. Rather, it is assumed that meanings are produced and exchanged in interpretative processes and that research has to begin with analyzing the concepts produced and used in these processes.

For example, the interpretative paradigm includes the three simple premises of symbolic interactionism. Blumer summarizes the basic assumptions of symbolic interactionism as 'three simple premises', that people 'act toward things on the basis of the meanings that the things have for them ... that the meaning ... is derived from, or arises out of, the social interaction ... with one's fellows ... and that these meanings are handled in, and modified through, an interpretative process used by the person in dealing with the things he encounters' (1969, p. 2). The consequence of these premises is that the different ways, that is how subjects ascribe meaning to objects, events, experiences, etc., become the central starting points of research. Reconstructing such subjective viewpoints becomes an instrument for analyzing social worlds. Another basic assumption has been formulated in the so-called 'Thomas-Theorem', which

> claim[s] that when a person defines a situation as real, this situation is real in its consequences, leads directly to the fundamental methodological principle of symbolic interactionism: researchers have to see the world from the angle of the subjects they study. (Stryker 1976, p. 259)

A similar point applies to rules of activities. These rules, too, are not unambiguously defined and are not applied by people in the same way (which the researchers then could take as given). Rather, they are defined by the participants in mundane processes of nego-tiation and interpretation. To study these processes of negotiation is a central approach in research, according to the interpretative paradigm.

An understanding of knowledge and science, which is the background of the interpre-tative paradigm, is the basis for the different approaches in qualitative research, which take the interpretation of subjective views or the detailed analysis of everyday practices as starting points for the study of research issues. The testing of hypotheses plays no big role in this context. The scientists' skepticism here is not limited to their own concepts (hypotheses or theories), which have to be tested empirically, but is extended to under-standing on the part of participants. Accordingly, you can no longer simply assume that participants share the same meanings. Rather, you should first clarify which (and which different) understandings of an issue or concept (e.g. 'often feeling under pressure' in our example above) are used in the participants' everyday lives and how different they are for various participants.

Constructionism

An alternative to the epistemological positions mentioned so far is social constructionism (see Flick 2004b for further detail of what follows).

A number of programs with different starting points are subsumed under this label. What is common to all constructionist approaches is that they examine the relationship to reality by using constructive processes to approach it. Constructionism is not a unified program, but develops in parallel fashion in a number of disciplines: psychology, sociol-ogy, philosophy, etc. In qualitative research in particular, this has a major influence, with the assumption that the realities we study are social achievements produced by actors, interactions and in situations. In the following approaches, we can see how the develop-ment of knowledge and its functions can be described in a constructionist way.

Social constructionism

Social constructionism in the tradition of Schütz (1962) or Berger and Luckmann (1966) asks for social (e.g. cultural or historical) conventionalization, which influences perception and knowledge in everyday life. Thus, Schütz (1962, p. 5) takes the following premise as a starting point: 'All our knowledge of the world, in common-sense as well as in scientific thinking, involves constructs, i.e. a set of abstractions, generalizations, formalizations and idealizations, specific to the relevant level of thought organization'. For Schütz, every form of knowledge is constructed by selection and structuring. The individual forms differ according to the degree of structuring and idealization, and this depends on their functions – more concrete as the basis of everyday action or more abstract as a model in the construction of scientific theories. Schütz enumerates different processes which have in common that the formation of knowledge of the world is not to be understood as the simple portrayal of given facts, but rather the contents of knowledge are constructed in a process of active production. Facts only become relevant through their meanings and interpretations: 'Strictly speaking, there are no such things as facts, pure and simple. All facts are from the outset facts selected from a universal context by the activities of our mind' (1962, p. 5). This also means that immediate access to experience is impossible. Rather, facts are 'always interpreted facts, either facts looked at as detached from their context by an artificial abstraction or facts considered in their particular setting. In either case, they carry their interpretational inner and outer horizons' (1962, p. 5).

A main idea is the distinction that Schütz makes between constructs of the first and second degree: 'the constructs of social science are, so to speak, constructs of the second degree, that is, constructs of the constructs made by the actors on the social scene'. Accordingly, Schütz holds that 'the exploration of the general principles according to which man in daily life organizes his experiences, and especially those of the social world, is the first task of the methodology of the social sciences' (1962, p. 59).

In particular, social science research is confronted with the problem that it encounters the world only through those versions of this world which subjects construct through interaction. Scientific knowledge and presentations of interrelations include different processes of constructing reality. For social constructionism, the social processes of exchange have a specific importance in the development of knowledge, particularly of the concepts that are used. Gergen (1994) formulates a number of assumptions for a social constructionism: in using these assumptions, Gergen highlights that terms for analyzing the world are formulated by researchers: they do not simply stem from what is analyzed within them. He highlights that these terms are constructed in communicative processes among people and may change over time. The construction of terminology is both a linguistic and a social phenomenon: language itself plays a central role in setting up the terms used by researchers but they (the researchers) are also influenced by use of these terms in relationships: 'Language derives its significance in human affairs from the way in which it functions within patterns of relationship' (Gergen 1994, pp. 49–50).

Common to the various social constructionist approaches is that they extend the skepticism of the epistemological programs mentioned above one step further: not only is the critical assessment of the scientist's theories (as in critical rationalism) and of the development and use of concepts (as in the interpretative paradigm) demanded. Rather, the processes of developing concepts on both sides – the participants and the scientists – are critically put into question and it is necessary to understand and analyze these processes as processes of construction. It becomes evident in this context that such processes of construction also determine scientific knowledge in the laboratories of the natural sciences.

Figure 2.1 summarizes the worldviews that have been discussed so far. In the bottom half of the figure, we find the worldviews of critical realism and positivism which start from seeing the phenomenon on a given ground as assumed in the normative paradigm of research. We can measure what we study and assume that participants and researchers share the same everyday knowledge about the phenomenon and the rules of ascribing meaning to it, for example. Above the fold, we find approaches that assume that interpretations and constructions of the phenomenon are important and vary between the researchers and the participants. These approaches are part of the interpretive paradigm and include the various versions of (social) constructionism.

Social Research between Fundamentalism and Pragmatism

You cannot simply choose one of the epistemological approaches presented in this chapter for your study as you would decide for or against a specific method (see Chapter 8). Rather, an epistemological position is the starting point for conceptualizing a research problem and for selecting the methods used to study it. At the same time, methodological approaches are grounded in epistemological positions. A quantitative study mostly aims at testing hypotheses formulated beforehand and navigates in the framework of critical rationalism, sometimes of positivism. Many qualitative studies with interviews focus on analyzing subjective experiences and views and are mostly done in the context of the interpretative paradigm. As a social researcher, you should be aware of these links when deciding on one or the other methodological approach and familiarize yourself with the background assumptions in each of the approaches.

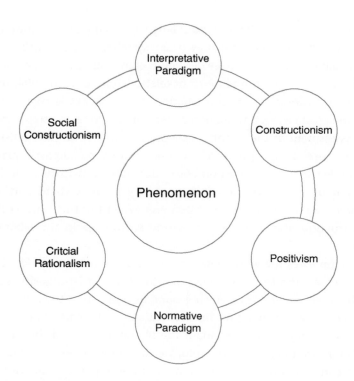

Figure 2.1 Worldviews, paradigms and phenomena

Epistemology has three functions in the context of social research:

- To provide a theoretical fundament for conceptualizing the research and procedure.
- To provide a basis for critical reflection on methods and the knowledge they produce.
- To support, improve and make social research possible and realizable.

Advantages and limits of the approaches presented in this chapter can be summarized as follows:

Critical rationalism provides a fundament for a major part of quantitative research. However, this epistemological fundament is mostly not further reflected on or questioned. Studies are pursued in applying methodological standards and rules. Hypotheses are formulated and tested. A wide range of data (in surveys and statistics, for example) with representative results referring to the population is produced. This research is based on the tacit assumptions mentioned above. The advantage is that these assumptions and most practices of standardization, for example, are no longer put into question (nor is there seen to be any need to do so). Methods for surveys (in sociology) or for experiments (in psychology) are available and can be applied like a toolkit. Thus, procedures are marked by a high degree of pragmatism. Limits are that the context of discovery, the views of participants in studies and qualitative methods, which are not testing hypotheses but address these views, remain widely excluded or limited to preliminary studies. Knowledge produced in this context focuses on assessing what is known or assumed – what could be formulated in a hypothesis beforehand. To discover something really new does not play a major role as a goal.

The interpretative paradigm provides a fundament for a major part of qualitative research. Here, the epistemological background assumptions are more extensively reflected in doing the actual research: How to give enough room to participants' views? How to document and analyze the making of social practices in enough detail? Here, analyzing a limited number of cases and fields includes exact and detailed descriptions that are further condensed into patterns and typologies. The advantages are that research refrains from tacit and implicit assumptions and pursues the aim of discovering something new. For example: What do the participants understand by XYZ? Application of the current variety of research methods for these purposes is better reflected here. A disadvantage is that the results pay for their exactness and detail with a limited scope, for example, in their generalization and statistical representativeness of the population. Also, they do not fulfill claims for secure knowledge in the way formulated by critical rationalism.

The example in Box 2.1 demonstrates how a student project with qualitative methods addressed a research question for which not much empirical research was available. Here the concept of social constructionism was applied.

Box 2.1   Research in the real world

A student's research study on children of members of a one-percenter motorcycle club

In her master's thesis, Alina Kaulmann (2017) studied how the moral socialization of children was affected by their fathers' membership of a specific motorcycle club (a one-percenter). She interviewed six now young adults whose fathers had been members of the Hell's Angels for years. Kaulmann used episodic interviews (see Chapter 11) and analyzed them using grounded theory coding (see Chapter 12). The interviews focused on children's relations to their father, their experience with

(Continued)

the club, the perception of the club within society, contact made with the police, and the images of women and men held by the club and the children. The interviewees were aged between 16 and 29 years old (when interviewed), four female and two male, and mostly students at school or university. The interviews show the former children's double role in the family and in society, their experience of disadvantage and discrimination, and any remaining conflicts. The empirical basis consists of interviewees' experiences and reports about their views and concepts.

Constructionist approaches in the tradition of Schütz (1962) inform many of the methods and studies in the framework of the interpretative paradigm. Applying qualitative methods can contribute to analysis of the productive efforts of the actors, referring to specific contexts. The advantage is that epistemological backgrounds and fundaments are made explicit in this analysis.

The epistemological programs briefly discussed in this chapter define what researchers see as knowledge, what they accept as results and evidence and how they proceed in their research projects (see Chapter 7). Therefore, the term worldview is used for naming this kind of program. Such programs provide an orientation for how to evaluate other researchers' studies, which principles and aims the studies were based on and how the studies reached their aims.

What You Need to Ask Yourself

When you plan your empirical social research project, you should take the following questions into account and find an answer for them (see Box 2.2).

Box 2.2 What you need to ask yourself

Epistemology

1 Against which epistemological (and theoretical) background will you carry out your study?
2 How do you take this background into account in the study?
3 How do you use hypotheses in the empirical testing of theoretical assumptions?

4 How do you take into account that the testing of hypotheses should be oriented on the principle of falsification and not of verification?
5 If your study is oriented on the interpretative paradigm, how do you take participants' viewpoints into account and how do you incorporate them into the research, the data and the analysis?
6 Which methodological consequences follow for your study if you start from a constructionist epistemology, and how do you put these consequences in concrete terms?
7 How do the methods you choose fit the epistemological assumptions the study is based on?

You can also use these questions for assessing the existing studies of other researchers.

What You Need to Succeed

This chapter should have given you a rough orientation about the epistemologies in the field of social research. This should allow you to contextualize what the subsequent chapters will treat in more detail. This orientation should be a first step in helping you, later, to take on your own research and study on a more solid ground. The questions in Box 2.3 should help you to check what to take home from this chapter for the steps you will take next.

Box 2.3 What you need to succeed

Epistemology

1 Did I understand that there are several epistemological concepts of research?
2 Do I have an idea of the differences between the concepts discussed in this chapter?
3 Do I see that they influence what researchers do in a study in concrete terms?

What you have learned

- Social research can refer to a variety of epistemological backgrounds.
- These backgrounds differ in what they see as issues for research, in what are understood as 'facts' and in the role played by the testing of hypotheses.

(Continued)

- Qualitative research, in particular, takes the views, perspectives and interpretations of participants as starting points.
- Constructionist positions start by analyzing the mundane and scientific constructions of the issue under study and of reality as seen by the participants.
- Epistemology should provide a fundament for doing and reflecting on social research.

What's next

The following texts address epistemological issues from different angles for qualitative and quantitative research:

Bryman, A. (2016) *Social Research Methods*, 5th edn. Oxford: Oxford University Press.
Flick, U. (2018) *An Introduction to Qualitative Research*, 6th edn. London: Sage.

The following case study can be found in the online resources. It should give you an idea of how the concept of social construction is turned into a research question and a research project:

Edwards, D. (2014) '"I'd rather you ask me because I/I/I don't really know, you know": The Dilemma of (Auto)biographical Interviewing in Biographical Research', *SAGE Research Methods Cases*. doi: 10.4135/9781446273050135000014.

The following article can be found in the online resources as well. The authors link epistemological and ethical issues in planning their research and in reflecting the contact with their participants when using a number of qualitative methods:

Sörensson, E. and Kalman, H. (2018) 'Care and Concern in the Research Process: Meeting Ethical and Epistemological Challenges through Multiple Engagements and Dialogue with Research Subjects', *Qualitative Research*, 18(6): 706–21.

Research Methodology Navigator

Orientation

You are here in your project

- Why social research?
- Worldviews in social research
- Ethical issues in social research
- From research idea to research question

Planning and design

- Reading and reviewing the literature
- Steps in the research process
- Designing social research

Method selection

- Deciding on your methods
- Triangulation and mixed methods

Working with data

- Using existing data
- Collecting new data
- Analyzing data

Reflection and writing

- What is good research? Evaluating your research project
- Writing up research and using results

ETHICAL ISSUES IN SOCIAL RESEARCH

How this chapter will help you

You will:

- see how ethical issues are involved in social research projects
- develop your sensitivity to ethical questions in social research
- appreciate the complexity of ethical considerations, and
- be able to plan and conduct your research project within an ethical framework.

In later chapters, we will consider the strengths and limitations of social research. There the main focus will be on methodological or technical limitations. We will consider in more detail such questions as: What can we grasp with one method, what is missed by it, and how can we overcome this by using several methods? We will also consider a more fundamental limitation, asking: When should you refrain from doing your research?

This chapter, however, focuses on limitations to social research to keep in mind, though of a different sort. We explore such questions as: Which ethical problems should be taken into account in research? Which ethical boundaries are touched and how can you approach ethical issues in doing your social research project? As we shall see, these questions involve us in some very general rules and problems.

Principles of Ethically Acceptable Research
Definitions of research ethics

Ethical issues are relevant to research in general. They are especially relevant in medical and nursing research. Here we find the following definition of research ethics, which may be applicable to other research areas too:

> Research ethics addresses the question of which ethically relevant issues caused by the intervention of researchers can be expected to impact on the people with or about whom they research. It is concerned in addition with the steps taken to protect those who participate in the research, if this is necessary. (Schnell and Heinritz 2006, p. 17)

Principles

In the context of the social sciences, Murphy and Dingwall (2001, p. 339) have developed an 'ethical theory' that provides a useful framework for this chapter. They discuss: *non-maleficence*, which means that researchers should avoid any harm to people resulting from participating in a study; *beneficence*, which implies that research involving human subjects should not simply be carried out for its own sake but should also promise or produce some positive and identifiable benefit; *autonomy* or *self-determination*, meaning that research participants' values and decisions should be respected; and *justice*, which means that all people should be treated equally in the study. Thus, the protection of participants and reflection on the necessity and usefulness of a study determine such a theory. We shall examine in more detail below how these principles apply in social research.

Schnell and Heinritz (2006, pp. 21–4), working in the context of health sciences, have developed a set of eight principles specifically concerning the ethics of research. These principles are as follows. Researchers should be able to: (1) justify why their research is necessary, (2) state what its aims are and (3) the circumstances under which subjects participate. Researchers should also (4) be able to make their procedures comprehensible to non-experts, and be able to (5) estimate the positive or negative consequences for participants, as well as (6) the possible violations and damage and (7) how to prevent these. (8) False statements about the usefulness of the research have to be avoided and data protection rules should be respected. These principles are influenced by the context of health research in which they were formulated but they can be transferred to guide other areas of social research.

Informed Consent

Informed consent as a general principle

It should be self-evident that studies should generally involve only people who (a) have been informed about being studied and (b) are participating voluntarily. Principles of informed consent and of voluntary participation for social research are to be found in the code of ethics of the German Sociological Association. For example:

> A general rule for participation in sociological investigations is that it is voluntary and that it takes place on the basis of the fullest possible information about the goals and methods of the particular piece of research. The principle of informed consent cannot always be applied in practice, for instance if comprehensive pre-information would distort the results of the research in an unjustifiable way. In such cases an attempt must be made to use other possible modes of informed consent. (Ethik-Kodex 1993: IB2)

Overall, such principles aim to ensure that researchers are able to make their procedures transparent (necessity, aims, methods of the study), that they can avoid or eliminate any harm or deception for participants, and that they take care of data protection.

There are of course some difficulties here. As mentioned in the quote, informing participants beforehand may contradict the aims of a study. Also, there are research settings in which it is not possible to inform all people who might become part of the research in advance. For example, in observations in open spaces (marketplaces, train stations, etc.), many people just passing by might become part of the observation for very short moments. For these people, it will be very difficult to obtain their consent. If, however, this is not the case, and consent *can* practically be obtained, you should never refrain from doing so. Accordingly, it is generally assumed that informed consent is a precondition for participating in research. To apply this principle in concrete terms, you can find some criteria in the literature:

- The consent should be given by someone competent to do so.
- The person giving the consent should be adequately informed.
- The consent should be given voluntarily. (Allmark 2002, p. 13)

Informed consent in researching vulnerable groups

Research with people who are, for special reasons, unable to give their consent raises specific ethical problems. These people are termed vulnerable groups. The term 'vulnerable subjects' refers to 'people who, because of their age or their limited cognitive abilities, cannot give their informed consent or who, because of their specific situation, would be particularly stressed or even endangered by their participation in a research project' (Schnell and Heinritz 2006, p. 43).

How then to proceed if you want to do research with people not able or not seen as able to understand your concrete procedures or to assess them and to decide independently? Examples include small children and people who are very old or who suffer from dementia or mental problems (for research with vulnerable people, see Liamputtong 2007). In such cases, you could ask other people to give the consent as substitutes – children's parents, family members or responsible medical or nursing staff in the case of the aged or ill. But do you then still meet the criteria of informed consent? Can you always assume that these other persons will take the same perspective as the participants you want to study? If you apply the principle of informed consent in a very strict sense in such cases, research is not permitted with these groups of participants, and thus research about relevant issues from the viewpoint of those concerned would be forgone. If you are conducting research involving vulnerable people, you should certainly not ignore the principle of informed consent. You should establish a way in which informed consent can be obtained either from or for the participants: consider carefully who is able to give this consent either together with or for them. However, these are trade-offs that can only be made for a specific study and not generally. There is no general rule on how to manage this problem: you will have to think for your particular study and your particular target group about how to solve this dilemma between doing necessary research and avoiding any mistreatment of your participants.

Box 3.1   Research in the real world

A prominent example of ethics gone wrong

In the 1960s, Humphreys (see 1975) conducted an observational study of homosexuals' sexual behavior. This study is an example of what can be done wrong in a study on the level of research ethics. It also stimulated a debate on the ethical problems of observations in fields like this. This debate continued for a long time, because it made visible the dilemmas of non-participant observation (see Chapter 11).

Humphreys did observations in public restrooms, which were meeting places in the gay subculture. As homosexuality was still illegal at that time, toilets offered one of the few possibilities for clandestine meetings. This study is an example of observation without participation, because Humphreys conducted his observation explicitly from the position of sociological voyeur, not as a member of the observed events and not accepted as an observer. In order to do this, Humphreys took on the role of somebody (the 'watch-queen') whose job it was to ensure that no strangers approached the events. In this role, he could observe all that was happening without being perceived as interfering and without having to take part in the events. After covertly observing the practices in the field, Humphreys then went on to collect participants' car licence plate numbers and used this information to obtain their names and addresses without informing the participants. He used this information to invite a sample of these members to take part in an interview survey.

Humphreys used unethical strategies to disclose participants' personal information in what was originally an anonymous event. At the same time, he did much to keep his own identity and role as a researcher concealed by conducting covert observation in his watch-queen role. Each aspect of such conduct is unethical in itself – keeping the research participants uninformed about the research and violating the privacy and secrecy of the participants.

The example in Box 3.1 may be a particular case, but it highlights several dilemmas in doing research in an unethically sound way. It may not be the most current example of research, but, as a more recent article demonstrates (Calvey 2019), the idea of covert research and participation is subject to 'rehabilitation'. The more general question is: What does research ethics mean and what would be the implications for doing research?

Confidentiality, Anonymity and Data Protection

Box 3.2 provides an example of a form developed for the author's own research projects. It is helpful to consider it here as a way to focus both on the issues of informed consent (discussed above) and on anonymity and data protection for participants.

The form should be completed and signed by both the researcher and the participant. Sometimes an oral agreement can be used as a substitute for the written contract if the participant does not want to sign it. Note that the form specifies a certain period after which participants can withdraw their consent. Furthermore, the form specifies who will have access to the data and whether or not the data can be used for teaching after anonymization.

Box 3.2 Research in the real world

Agreement about data protection for scientific interviews

- Participation in the interview is voluntary. It has the following purpose:

  [issue of the study]

  Responsible for doing the interview and analyzing it are:

  Interviewer:

  [name]

  [name of institution]

  Supervisor of the project:

  [name]

  [name and address of institution]

  The responsible persons will ensure that all data will be treated confidentially and only for the purpose agreed herewith.

- The interviewee agrees that the interview will be recorded and scientifically analyzed. After finishing the recording, (s)he can ask for erasure of particular parts of the interview from the recording.

- For assuring data protection, the following agreements are made (please delete what is not accepted):

  The material will be processed according to the following agreement about data protection:

Recording

1   The recording of the interview will be stored in a locked cabinet and in password-protected storage media by the interviewers or supervisors and erased after the end of the study or after two years at the latest.

2   Only the interviewer and members of the project team will have access to the recording for analyzing the data.

3   In addition, the recording can be used for teaching purposes. (All participants in the seminar will be obliged to maintain data protection.)

Analysis and archiving

1   For the analysis, the recording will be transcribed. Names and locations mentioned by the interviewee will be anonymized in the transcript – as far as necessary.

2   In publications, it is guaranteed that identification of the interviewee will not be possible.

• The interviewer or the supervisor of the project holds the copyright for the interviews.

• The interviewee may revoke his or her declaration of consent completely or partially within 14 days.

[Location, date]: ....

Interviewer: .... Interviewee: ....

In the case of an oral agreement

I confirm that I informed the interviewee about the purpose of the data collection, explained the details of this agreement about data protection, and obtained his or her agreement.

• [Location, date]: .... Interviewer: ....

Confidentiality and anonymity may be particularly relevant if the research involves several participants in a specific, very small, setting. If you interview employees in the same enterprise or family members independently of each other, it will be necessary to ensure confidentiality not only with respect to the public beyond that setting, but also within it. Readers of a publication should not be able to identify the individuals who participated as interviewees, for example. Therefore, you should change personal data such as names, addresses, workplaces, etc. so that inferences to persons and suchlike become impossible or, at the very least, are hampered. Accordingly, the researcher has to ensure that other participants cannot identify their colleagues in the presentation of their common workplace or in what the researcher reveals about their study. For this purpose, a consistent anonymization of the data and a parsimonious use of context information are necessary.

If children are interviewed, parents often want to know what their children said in the interviews – which may be problematic for interviews referring to relations of parents to children or to conflicts among them. To avoid this problem, it may be necessary to inform the parents beforehand when such information cannot be given to them.

It is particularly important to store the data (questionnaires, recordings, transcripts, field notes, interpretations, etc.) physically in a safe and locked way (e.g. data safes, cupboards that can be locked), so that no one gains access to the data who is not supposed to have this access (see Lüders 2004b). The same precautions must be taken if the data are stored electronically – which means that they are password protected at least, and that the number of persons having access to the site is strictly limited.

How to Avoid Causing Harm to Participants

The risk of harm for participants is a major ethical issue in social research. If, for example, you ask in an interview or a questionnaire how people live with their chronic illness and cope with it, you will confront your respondents with the severity of their illness and maybe with the limits to or lack of their life expectancy, again or additionally. This might cause a crisis or lead to extra stress for the interviewees. Is it ethically correct to produce such a risk for the participants in the research?

Box 3.3 Research in the real world

A student's research collecting data from a vulnerable group of older people

Wolfram Herrmann (see Herrmann and Flick 2012) did his PhD in the context of a larger study on residents' sleep disorders in nursing homes. His study focused on the residents' views of the issue of sleep, their subjective perceptions and subjective explanation of why they did not sleep well, and on how they dealt with this problem. Herrmann interviewed 30 residents in five nursing homes and used the episodic interview as a method of data collection, after the research was approved by the ethics committee of a large medical university in Berlin. The ethical issues concerned how to obtain the informed consent of the potential interviewees, and how to find out whether they were (still) able to give such consent and to give an interview. As a large part of the nursing home population in Germany suffers from cognitive limitations, the participants had to be selected according to criteria as to whether they were sufficiently oriented in 'person and place' to be included in such interviews. Ignoring such a selection criterion would have meant conducting interviews with people unable to give informed consent, while applying it would exclude a major group of residents (and potential interviewees) beforehand. Taking such a criterion seriously may limit the application of results to the target group in general (nursing home residents), while ignoring the criterion may make the quality of the data questionable, not only on an ethical but also on a methodological level. Being confronted with the need to apply such a selection criterion may also produce an ethical dilemma in the step of sampling (see Chapter 7).

A specific problem arises in testing the effects of medications (or other forms of inter-vention) in randomized control studies (see Chapters 7 and 13). Here, people with a diagnosis are randomly allocated to an intervention group (receiving treatment with the medication) or to a control group (receiving instead a placebo without effect, which means no treatment). Is it ethically justified to deprive this second group of a treatment, or to give it to them only after the end of the study? Should you do randomized studies in such a case, particularly if it is about a serious or life-threatening illness? (See Thomson et al. 2004 on this issue.)

The examples just given are drawn from medical research. However, the need to avoid harm applies to *all* research, not just medical studies. According to the code of ethics of the German Sociological Association:

> Persons who are observed, questioned or involved in some other way in investigations, for example in connection with the analysis of personal documents, shall not be subject to any disadvantages or dangers as a result of the research. All risks that exceed what is normal in everyday life must be explained to the parties concerned. The anonymity of interviewees or informants must be protected. (Ethik-Kodex 1993: IB5)

Demands on participants resulting from the research

Research projects always make demands on participants (see Wolff 2004). For example, participants may be required to sacrifice time to complete a questionnaire or to answer the interviewers' questions. In addition, they may be expected to deal with embarrassing questions and issues and to give the researchers access to their privacy.

From an ethical point of view, you should reflect on whether the demands your research would make on participants are reasonable – especially in the light of their specific situations. For example, you should consider whether a confrontation with their own life history or illness in an interview or survey might even intensify their situation.

Rule of economy of demands and stress

If you are requesting personal information, you should always consider whether you really need a whole life history (e.g. in a narrative interview) for answering your research question, or whether responses to more focused questions might be sufficient. However, we can observe a trend in quantitative research to add this questionnaire to the data collection or to include that question in the survey. This leads, on the one hand, to an extension of the datasets in the single study. On the other hand, it can be seen as a demonstration that questions of economy are not only an issue for qualitative research. In both contexts, you should check what are justifiable demands on participants, what is already stressful and no longer justified as a demand, and when harming the participants begins.

Codes of Ethics

Many scientific associations have published codes of ethics. They are formulated to regulate the relations between researchers and the people and fields they study. Sometimes they also regulate how therapists or caregivers should work with their clients or patients, as in psychology and nursing. Some of them refer to specific questions of the research in the area, as in research with children in education. Examples of codes of scientific associations, available on the Internet, include the following (all last accessed 23 April 2019):

- The British Educational Research Association [BERA] has published a fourth edition of *Ethical Guidelines for Educational Research* (www.bera.ac.uk/researchers-resources/publications/ethical-guidelines-for-educational-research-2018).
- The British Psychological Society (BPS) has published a Code of Conduct, Ethical Principles and Guidelines (www.bps.org.uk/news-and-policy/bps-code-ethics-and-conduct).
- The British Sociological Association (BSA) has formulated a Statement of Ethical Practice (www.britsoc.co.uk/media/24310/bsa_statement_of_ethical_practice.pdf).
- The American Sociological Association (ASA) refers to its Code of Ethics (www.asanet.org/code-ethics).
- The Social Research Association (SRA) has formulated Ethical Guidelines (www.the-sra.org.uk/SRA/Ethics/Research-ethics-guidance/SRA/Ethics/Research-Ethics-Guidance.aspx?hkey=5e809828-fb49-42be-a17e-c95d6cc72da1).
- The German Sociological Association (GSA) has developed a Code of Ethics (www.soziologie.de/index_english.htm).

Such codes of ethics demand that research be pursued only under the condition of informed consent and without harming the participants. This includes a requirement that the research does not intrude on participants' privacy in inappropriate ways and that participants are not deceived about the research aims.

Ethics Committees

Professional associations, hospitals and universities typically have ethics committees to ensure that ethical standards are met:

> Ethics committees are in charge of assessing whether researchers have made enough ethical considerations before beginning the research they plan. For this purpose, ethics committees have two instruments. They can decide about projects by accepting or rejecting them. Secondly they can become active in consulting researchers and discuss with them suggestions for the ethical planning of a project. (Schnell and Heinritz 2006, p. 18)

For this purpose, the committees assess proposed research designs and methods before they are applied to human beings. Such assessments normally consider three aspects (see Allmark 2002, p. 9): (a) scientific quality; (b) the welfare of participants; and (c) respect for the dignity and rights of participants.

A relevant question for ethics committees is whether or not a research project will provide new insights to add to existing knowledge. A project that merely duplicates earlier results can be seen as unethical – in particular, research that repeatedly does the same studies again (see e.g. Department of Health 2001). Here, the question is raised of how the stress for participants is justified by the benefits to science and the novelty of results. Exceptions are studies with the explicit aim of testing whether it is possible to replicate findings from earlier studies.

In considering the quality of research, we can see a source of conflict. To be able to judge the quality of research, the members of the ethics committee should have the necessary knowledge to assess a research proposal on a methodological level. In effect, this may mean that members of the committee, or at least some of the members, should be researchers themselves. Yet, if you talk for a while with researchers about their experiences with ethics committees and with proposals submitted to them, you will come across many stories about how a research proposal was rejected because the members did not understand its premise, or lacked the methodological background of the applicant, or simply disliked the style of research. Thus, ethics committees may in practice end up rejecting research proposals for non-ethical reasons. Such a reservation can be particularly strong when a qualitative research proposal is confronted by committees or members who think only in natural science terms, or where experimental research is confronted by committees mainly concerned with interpretative categories.

In assessments by ethics committees, questions of welfare often involve weighing the risks (for participants) against the benefits (of new knowledge and insights about a problem or of finding a new solution to an existing problem). For example, if you want to find out the effects of a medication in a control group study (see Chapter 7), you will give participants in the control group a placebo rather than the medicine under study. If the control group is to be comparable to the study group, you will need people for it who are also in need of treatment with the medication. Thus, the dilemma arises between depriving the control group members of a possible treatment (at least for the moment) and otherwise being unable to study the effects of the medication adequately (see Thomson et al. 2004). Again, we find a potential conflict here: weighing the risks and benefits is often relative rather than absolute and clear.

The dignity and rights of the participants pertain to issues of (a) consent given by the participant, (b) sufficient information provided as a basis for giving consent, and (c) the need for consent to be voluntary (Allmark 2002, p. 13). Beyond this, researchers need to guarantee participants' confidentiality. This requires that the information about them will be used only in ways that make it impossible for other persons to identify them or for any institution to use it against the interests of the participants.

Ethics committees review and canonize the principles discussed here. For a detailed discussion of such principles, see Hopf (2004) and Murphy and Dingwall (2001).

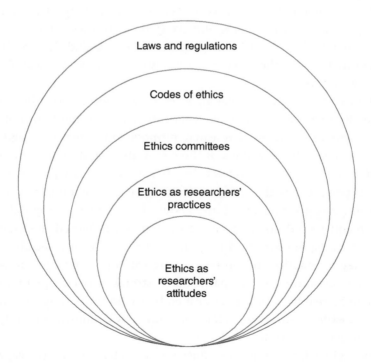

Figure 3.1 Contexts and levels of research ethics

In Figure 3.1, we find the levels and contexts relevant for research ethics. The first level is the attitude with which researchers approach their study and how far it is informed by ethical attitudes. This attitude sets the ground for the researchers' practices in the field and how they are driven and limited by ethical considerations. Ethics committees assessing a proposal define the context of a research study and delimit what is acceptable as practices with respect to research ethics from an external viewpoint in the concrete study. Codes of ethics define a framework for ethically sound research and practices in a specific field, for example a profession or a discipline. Laws and regulations in a country (or in the European Union, for example) define the legal basis and context for research ethics. Laws and regulations are the most general context, which becomes relevant through the other levels for the researchers' attitudes and practices.

Rules of Good Scientific Practice

Unfortunately, researchers have sometimes been guilty of concealing their results (for examples, see Black 2006). Because of this, the German Research Council has developed proposals for safeguarding good scientific practice. These are outlined in Box 3.4 (Deutsche Forschungsgemeinschaft, DFG 2013). These rules define standards concerning honesty in using data, scientific fraud and the documentation of original data (completed questionnaires, recordings and transcripts of interviews, etc.).

Research Ethics: Cases and Mass Research

Ethical principles in social research apply to qualitative as well as quantitative research – even though the concrete questions and details involved may be very different. Data protection and anonymization may be more easily guaranteed for a single participant in the random sampling and statistical analysis of data than for a participant in a qualitative study with purposive sampling and (a few) expert interviews. Note, in particular, that if you plan a survey with repeated data collection from the same people in the first instance, you will need to store the real contact data of the participants so that you can return to them. There is a risk here of inadvertently infringing participants' rights to anonymity and data protection. Contact data and questionnaire responses will need to be separated.

Research Ethics in Digital and Online Research

Gaiser and Schreiner (2009, p. 14) have listed a number of questions to consider from an ethical point of view if you are planning an online study. They are:

- Can participant security be guaranteed? Anonymity? Protection of the data?
- Can someone ever really be anonymous online? And if not, how might this impact on the overall study design?
- Can someone 'see' a participant's information, when s/he participates?
- Can someone unassociated with the study access data on a hard drive?
- Should there be informed consent to participate? If so, how might online security issues impact on the informed consent?
- If a study design calls for participant observation, is it OK to 'lurk'? Is it always OK? If not, then when? What are the determining factors?
- Is it ever OK to deceive online? What constitutes online deception?

This list shows how general issues of research ethics are relevant for online as well as traditional research.

Box 3.4   Research in the real world

Rules for good scientific practice

Recommendation 1

Rules of good scientific practice shall include principles for the following matters (in general, and specified for individual disciplines as necessary):

- fundamentals of scientific work, such as:
  - observing professional standards
  - documenting results
  - consistently questioning one's own findings
  - practising strict honesty with regard to the contributions of partners, competitors and predecessors
  - cooperation and leadership responsibility in working groups
  - mentorship for young scientists and scholars
  - securing and storing primary data (recommendation 7)
  - scientific publications (recommendation 11).

Recommendation 7

Primary data as the basis for publications shall be securely stored for ten years in a durable form in the institution of their origin.

Recommendation 8

Universities and research institutes shall establish procedures for dealing with allegations of scientific misconduct. They must be approved by the responsible corporate body. Taking account of relevant legal regulations, including the law on disciplinary actions, they should include the following elements:

- a definition of categories of action which seriously deviate from good scientific practice (recommendation 1) and are held to be scientific misconduct, for instance the fabrication and falsification of data, plagiarism or breach of confidence as a reviewer or superior
- jurisdiction, rules of procedure (including rules for the burden of proof) and time limits for inquiries and investigations conducted to ascertain the facts
- the rights of the involved parties to be heard, and to discretion, and rules for the exclusion of conflicts of interest
- sanctions depending on the seriousness of proven misconduct
- jurisdiction for determining sanctions.

Recommendation 11

Authors of scientific publications are always jointly responsible for their content. A so-called 'honorary authorship' is inadmissible.

Source: excerpt, Deutsche Forschungsgemeinschaft 2013

Data Protection: Regulations in the European Union

Data protection has become a major concern in Europe in the context of social media and Big Data (see Chapter 10) and in the use of both mainly for commercial but also for research interests. This is not only an issue in the context of research – the European Union (EU) has passed new regulations for data protection and has discussed this as an issue of ethics in research and beyond: 'Data protection is both a central issue for research ethics in Europe and a fundamental human right' (EU 2018, p. 3). The regulations passed the EU parliament in 2016 and have implications for research in the EU, in every country in the EU and beyond, if countries or subjects from the EU are involved. For research, the EU (2018) has published a document on ethics and data protection, which was developed by a panel of experts, and underlines the notion that in

> research settings, data protection imposes obligations on researchers to provide research subjects with detailed information about what will happen to the personal data that they collect. It also requires the organisations processing the

data to ensure the data are properly protected, minimised, and destroyed when
no longer needed. (EU 2018, p. 3)

This document highlights the responsibilities of the researchers or their institutions to provide sufficient information to participants as a consequence of the EU's 2016 General Data Protection Regulation (GDPR). The focus is on data whose processing entails a *high risk* for participants, because of the kinds of data or information that are being used (e.g. political opinion, religious beliefs), due to the specific subjects involved (e.g. children, vulnerable people, people who have not given their consent), due to the specific data collection or processing techniques (which are privacy-invasive, for example, or based on data mining), or due to the involvement of non-EU countries (see 2018, p. 6). In such cases of higher-risk data processing, researchers have to *provide a detailed analysis* of ethics issues raised by their project, including:

- an overview of all planned data collection and processing operations
- an identification and analysis of the ethics issues that these raise, and
- an explanation of how you will mitigate these issues in practice. (EU 2018, p. 7)

Data protection revolves around three concepts: references to research subjects should be based on *pseudonymization*, in which all personal information (e.g. individuals' names) is substituted by unique identifiers that are not connected to their real-world identity; *anonymization* converts personal data into anonymized data without identifiers; and *re-identification* is a process which turns pseudonymized or anonymized data back into personal data (EU 2018, p. 8). This may be necessary if you want to do a longitudinal study with the same participants and need to contact them again (see Chapter 7). Otherwise, this is discussed as a risk to minimize or avoid – in the sense that other people might be able to re-identify participants of a study. A further issue is the need to restrict the data that are collected to what is absolutely necessary: 'Data processing must be lawful, fair and transparent. It should involve only data that are necessary and proportionate to achieve the specific task or purpose for which they were collected (Article 5(1) GDPR)' (EU 2018, p. 10).

What are the key changes in the GDPR as compared to earlier regulations and practices? A first change is the extended territorial scope in that the regulations also now apply to companies and researchers from outside the EU. If the research is in another country but involves EU citizens or, vice versa, if EU researchers work with people outside the EU, the GDPR applies as well. Other changes concern the (high) penalties for breaching GDPR rules that have been defined, the new rights for data subjects – such as having to be notified in case of a breach of GDPR regulations within 72 hours, the right to access one's own data and to be forgotten (i.e. to have data erased), and the privacy of participants now having to be a major concern for research design. Finally, research institutes should have

Data Protection Officers appointed according to their professional qualifications as staff members or external officers, who must be given the appropriate resources for protecting rights and data, and report to the highest levels of management (see www.trunomi.com).

The United Kingdom Research Institute (UKRI 2018) has published a helpful overview (*GDPR and Research: An Overview for Researchers*). It summarizes the lawful bases for data processing as set out in Article 6 of the GDPR for any research study. At least one of the following should be given to legitimize a study: (a) the individual's consent, (b) a contract with the individual, (c) a legal obligation to process the data, (d) vital interests (to protect someone's life), (e) a public task performed by the research, or (f) the legitimate interests of the researcher or a third party.

As a consequence, many universities have produced information packs to instruct their researchers on how to act according to the new EU regulations and following country- specific laws. For example, the University College of London (UCL 2018) has formulated a document called 'Guidance for Researchers on the Implications of the General Data Protection Regulation and the Data Protection Act 2018'. It summarizes the treatment of personal data, informed consent, data protection, fairness and transparency at EU and UK levels. When you plan your research, you can begin by consulting this guidance or, even better, check whether or not your university has a similar document and guidance available.

Figure 3.2 locates the ethical principles in research practice starting from informed consent as a pre-condition for avoiding harm for participants based on data protection and maintaining participants' privacy and leading to the exclusion of the deception of participants.

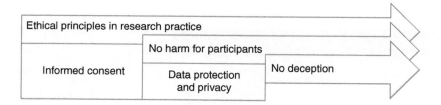

Figure 3.2 Ethical principles in practice

Conclusion

We may finish with two general points concerning ethical principles in research. First, we should remember that research entails questions of integrity and objectivity. As the German Sociological Association states:

Sociologists strive for scientific integrity and objectivity in pursuing their
profession. They are committed to the best standards that are possible in research,

teaching and other professional practices. If they make discipline specific judgments, they will represent their field of work, the state of knowledge, their disciplinary expertise, their methods and their experience unambiguously and appropriately. In presenting or publishing sociological insights, the results are presented without any biasing omission of important results. Details of the theories, methods and research designs that are important for assessing the results of the research and of the limits of their validity are reported to the best of the researchers' knowledge. Sociologists should mention all their funding sources in their publications. They guarantee that their findings are not biased by the specific interests of their sponsors. (Ethik-Kodex 1993: IA1–3)

This quotation refers specifically to 'sociologists'. It may be applied, however, to social researchers in general.

Second, questions of research ethics are raised in *any* kind of social research and with *all* sorts of methods. No methodological approach is free of ethical problems, even though these differ between methods: 'Research methods are not ethically neutral. That applies to qualitative, quantitative and triangulated methods in the same way. Criteria for assessing the quality of research at least implicitly ask for ethical issues as well' (Schnell and Heinritz 2006, p. 16).

What You Need to Ask Yourself

Box 3.5 What you need to ask yourself

Ethics

1 How will you put the principle of informed consent into practice?
2 Have you informed all participants that they are taking part in a study or are involved in it?
3 How will you ensure that the participants do not suffer any disadvantage or damage from the study or from taking part in it?
4 How will you make sure that the participants in a control group do not suffer any disadvantage from not receiving the intervention?
5 How will you guarantee the voluntariness of participation?
6 How will you ensure that vulnerable people (e.g. children or the cognitively impaired) have agreed to being interviewed, i.e. that not merely the consent of parents or caregivers was obtained?

7 How will you organize the anonymization of the data and how will you deal with issues of data protection in the study?

8 How will you take these issues into account for the storage of the data and in presenting the results?

9 Have you checked your method of proceeding against the relevant ethical code(s)?

10 If so, which problems became evident here?

11 Is a statement from an ethics committee necessary for your study and, if so, have you obtained it?

12 How will the project conform to the requirements formulated in this process?

13 What is the novelty in the expected results, which justifies carrying out your project?

14 Can you specify the expected results?

What You Need to Succeed

In pursuing an empirical project in social research, you should consider the ethical questions in Box 3.6. These questions can be applied to the planning of your own study and, in a similar way, to the evaluation of other researchers' existing studies.

Box 3.6   What you need to succeed

Ethics

1 Do I see the necessity of the principle of informed consent?

2 Do I understand what it means to protect participants from harm? Do I have an understanding of confidentiality and privacy?

3 Do I see the necessity and the practical issues of data protection?

4 Do I see the need for checking whether the existing research has already answered my research question?

5 Do I see the need for being skilled in applying the methods I have selected for my study?

What you have learned

- Every research project should be planned and assessed according to ethical principles.

(Continued)

- Voluntariness of participation, anonymity, data protection and avoidance of harm for participants are preconditions.
- Informed consent should be obtained for every research project. Exceptions to this have to be justified rigorously.
- Codes of ethics provide an orientation for taking ethical principles into account and applying them.
- Ethics committees seek to ensure that ethical principles are upheld.
- For qualitative and quantitative research, ethical questions may be raised in different ways.
- Ethical issues can arise at all stages of the research process – from creating an idea and a design to study it, to ways of disseminating and using the results.

What's next

Chapter 4 in the following book provides further discussion on the ethics of social research:

Flick, U. (2018) *An Introduction to Qualitative Research*, 6th edn. London: Sage.

The following chapter addresses specific issues of ethics in doing digital research:

Tiidenberg, K. (2018) 'Ethics in Digital Research', in U. Flick (ed.) *The SAGE Handbook of Qualitative Data Collection*. London: Sage. pp. 466–81.

The following chapter discusses specific issues of ethics in using data and findings in qualitative research:

Mertens, D.M. (2014) 'Ethical Use of Qualitative Data and Findings', in U. Flick (ed.), *The SAGE Handbook of Qualitative Data Analysis*. London: Sage. pp. 510–23.

The following case study can be found in the online resources. It shows how the issue of research ethics becomes and remains relevant throughout the research process and makes suggestions about what to do with it:

Lee, T. T. and Emmerich, N. (eds) (2019) 'Research Ethics as a Process: A Qualitative Study of Foreign Domestic Workers and their Employers in Hong Kong', *SAGE Research Methods Cases*. doi: 10.4135/9781526487001.

The following article can be found in the online resources. It critically discusses the limits of a basic principle of current research ethics:

Calvey, D. (2019) 'The Everyday World of Bouncers: A Rehabilitated Role for Covert Ethnography', *Qualitative Research*, 19(3): 247–62.Research Methodology Navigator

Research Methodology Navigator

You are here in your project

Orientation
- Why social research?
- Worldviews in social research
- Ethical issues in social research
- From research idea to research question

Planning and design
- Reading and reviewing the literature
- Steps in the research process
- Designing social research

Method selection
- Deciding on your methods
- Triangulation and mixed methods

Working with data
- Using existing data
- Collecting new data
- Analyzing data

Reflection and writing
- What is good research? Evaluating your research project
- Writing up research and using results

4

FROM RESEARCH IDEA TO RESEARCH QUESTION

—How this chapter will help you—

You will:

- recognize the starting points for social research
- appreciate where research questions come from
- understand how research questions differ between qualitative and quantitative research, and
- grasp the use of hypotheses.

This chapter seeks to show how research questions for empirical studies emerge from general interests and from the personal and social backgrounds of the researcher. For this purpose, let's look first at some examples.

Starting Points for Research

The literature of the history of social research recounts many examples of how ideas for research have emerged and been developed into research questions, but there are some which are particularly instructive for illustrating the origin of research questions.

For example, Marie Jahoda (1995; see also Fleck 2004, p. 59) has described the origins of her study with Paul Lazarsfeld and Hans Zeisel on *Marienthal: The Sociology of an Unemployed Community* (Jahoda et al. 1933/1971). The impulse for the study came in the late 1920s from Otto Bauer, leader of the Austrian Social Democratic Party. The background to the study included the Great Depression of 1929 and also the political interests and orientation of the researchers. As a result, the researchers developed the idea of studying how a community changes in response to mass unemployment. From this general idea, they formulated research questions concerning the attitude of the population towards unemployment and the social consequences of unemployment.

Another example, this time from the 1950s, is provided by Hollingshead and Redlich's (1958) study of social class and mental illness. Their study stemmed from the general observation that 'Americans prefer to avoid the two facts studied in this book: social class and mental illness' (1958, p. 3). From this starting point, they proceeded to explore possible relationships between social class and mental illness (and its treatment). For example, people with a low social status might be more at risk of becoming mentally ill and their chance of receiving good treatment for their illness might be lower than people with a high social status. From their general interest, the authors developed two research questions: '(1) Is mental illness related to class in our society? (2) Does a psychiatric patient's position in the status system affect how he [sic] is treated for his illness?' (1958, p. 10).

They then broke these two questions down into five working hypotheses (see below and 1958, p. 11):

1   The prevalence of treated mental illness is related significantly to an individual's position in the class structure.
2   The types of diagnosed psychiatric disorders are connected significantly to the class structure.
3   The kind of psychiatric treatment administered by psychiatrists is associated with the patient's position in the class structure.
4   Social and psychodynamic factors in the development of psychiatric disorders correlate with an individual's position in the class structure.
5   Mobility in the class structure is associated with the development of psychiatric difficulties.

To test these hypotheses, Hollingshead and Redlich conducted a community study in a city of 24,000 people. They included all psychiatric patients diagnosed within a certain period, using a questionnaire about their illness and their social status. They also interviewed health professionals.

A contrasting example, also from the mid-twentieth century, is provided by a study by Glaser and Strauss (1965). Following their experiences of their own mothers dying in hospital, they developed the idea of studying 'awareness of dying'. The authors (1965, pp. 286–7) described in some detail how these experiences stimulated their interest in the processes of communicating with and about dying persons and what they later described as 'awareness contexts'. Here, the background for developing the research idea, interest and question was very much a personal one – the recent autobiographical experiences of the researchers.

And – to give another example from a professional researcher – Hochschild (1983, p. ix) has described how early experiences, as a child, of her family's home and social life became the source of her later 'interest in how people manage emotions'. Her parents worked for the US Foreign Service. This provided Hochschild with opportunities to see and interpret the different forms of smiles – and their meanings – produced by diplomats from different cultural backgrounds. Hochschild learned from these experiences that emotional expressions, such as smiles and handshakes, conveyed messages on several levels – from person to person and also between the countries the people represented. This led, much later, to her specific research interest:

I wanted to discover what it is that we act upon. And so I decided to explore the idea that emotion functions as a messenger from the self, an agent that gives us an instant report on the connection between what we are seeing and what we had expected to see and tells us what we feel ready to do about it. (1983, p. x)

From that interest, she developed a study (*The Managed Heart*) of two types of public-contact workers (flight attendants and bill collectors), showing how work functioned to induce or suppress emotions when the workers were in contact with their clients.

The above examples are complemented by the student project detailed in Box 4.1.

Box 4.1   Research in the real world

Students' research on women's health concepts in Portugal and Germany

Two psychology students (Beate Hoose and Petra Sitta; see Flick, Hoose and Sitta 1998) developed an interest in the differences and similarities in what people understand as health in different contexts, for their diploma thesis. They decided, for personal reasons, to compare concepts of health (and illness) of women living in Berlin (as they did themselves) and in Lisbon (which one of them visited quite often, having

(Continued)

a good network there). By comparing the features of the health systems in Germany and Portugal, they developed clearer research questions focusing on the role of culture in health concepts. On a practical level, they did episodic interviews (see Chapter 11) with six women in each of the cultures and analyzed them by coding the material using grounded theory (see Chapter 12). This is an example of how two students developed their own research question and project starting from their own situation, everyday conversations and interests in practical issues. They found common factors in the backgrounds of their interviewees – a 'lack of awareness' in the Portuguese interviews and the feeling of being 'forced into health' in the German interviews.

If we compare the examples above, we can see that they show diverse sources for developing research interests, ideas and subsequently research questions. They range from very personal experiences (Glaser and Strauss 1965) to social experiences and circumstances (Hochschild 1983), through social observations (Hollingshead and Redlich 1958), to societal problems and political commissioning (Jahoda et al. 1933/1971), and to students' personal interest in an everyday issue (Flick et al. 1998). In each case, a general curiosity arose, which the researchers pursued and subsequently formulated in concrete terms.

Research, then, can have various starting points. In particular:

- Research problems are often discovered in everyday life. For example, in the everyday life of an institution someone may discover that, say, waiting times emerge in specific situations. In order to find out what determines waiting times and, perhaps, how they might be reduced, systematic research may be undertaken.
- Second, there may be a lack of data and empirical insights about a specific problem – for example, the health situation of young people in Germany – or about a specific subgroup, such as adolescents living on the street.
- A third source for identifying a research problem may be the literature. For example, a theory might have been developed which requires testing empirically. Or analysis of the existing literature may reveal that gaps exist in the knowledge about a problem. Empirical research may be designed to close such gaps.
- Fourth, research problems may grow out of previous studies producing new questions or leaving some questions unanswered.

Figure 4.1 outlines the process, starting from personal interests, informed by reading the literature and adopting a research perspective, leading to a research interest and becoming concrete in developing a research question.

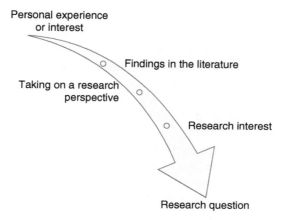

Figure 4.1   Developing a research question

Origins of Research Questions

We can illustrate the development of research questions by using two recent examples. They concern health behavior and the health of adolescents living on the street. The first example (see Box 4.2) comes from a quantitative study embedded in a worldview of critical rationalism. The second (see Box 4.3) comes from a qualitative study with a social constructionist worldview as background (see Chapter 2).

Box 4.2   Research in the real world

The 'Health Behavior in Social Context' study of the World Health Organization (WHO)

In the 'Health Behavior in Social Context' (HBSC) study, children and adolescents from 36 countries were interviewed, using a standardized questionnaire, concerning their health status and behavior. This research was conducted in order to produce a health report for the younger generation and thus to contribute to an improvement in illness prevention and health promotion for this age group. In Germany, for example, this survey has been repeatedly run since 1993 – most recently in 2010 (see Hurrelmann et al. 2003; Richter et al. 2008; Currie et al. 2012). Hurrelmann et al. have described the aims of the study as follows:

> In this youth health survey, several questions shall be answered: descriptive questions about physical, mental and social health and about health behavior; in

(Continued)

focus is the question of how far health-relevant lifestyles are linked to subjective health; and of how far personal and social risk and protective factors can be identified for the prevention of health problems together with their subjective representation in physical and mental respects. (2003, p. 2)

For this study in Germany, adolescents aged 11, 13 and 15 years were interviewed in schools. For the international study, a representative sample was drawn (see Chapter 7), comprising about 23,000 adolescents in different areas of Germany. For the German study, a subsample of 5,650 adolescents was drawn randomly from this sample. This subsample was asked to answer a questionnaire about their subjective health, risk of accidents and violence, the use of substances (tobacco, drugs and alcohol), eating, physical activity, peers and family, and school. The research question for this study resulted from an interest in developing a representative overview of the health situation and the health-relevant behavior of adolescents in Germany and in comparison with other countries.

Construction of research questions

Sandberg and Alvesson (2010) have addressed the issue of how research questions are constructed. They did an empirical study and found two major alternatives – gap-spotting or problematization. They analyzed 52 published studies in the field of organization research. They show that gap-spotting is a major source for formulating research questions for new studies. Gap-spotting means that gaps in the literature or in research are identified. Three basic modes of gap-spotting were found: 'confusion spotting' means that competing explanations for the same phenomenon co-exist and further empirical clarification is needed; 'neglect spotting' refers to overlooked areas in research, under-researched fields or a lack of empirical support for some explanations; 'application spotting' refers to an extension or a complementation of a theory to other fields. Sandberg and Alvesson see gap-spotting as the dominant way of identifying a research question for further studies, but see their second mode of identification as the more promising alternative. Problematization means that not only is a lack of research taken as a starting point, but there is also a questioning of what is taken for granted. This is the bridge between our last example and the following one. Whereas the HBSC study focuses on the average of the population – the health of children in various countries – the study in Box 4.3 questions whether such a broad approach can cover the specific health situation of marginalized groups. This latter study questions even more the assumption that we can simply apply general understandings of health to groups such as homeless adolescents.

Box 4.3   Research in the real world

Health on the street – homeless adolescents

A different approach to a similar topic is provided by our second example. The study in Box 4.2 provided a good overview of the health situation of the average young person in Germany and other countries. However, such a broad study cannot focus on particular (mainly very small) subgroups. The reasons for this are the use of random sampling and the fact that access to participants was via school. Adolescents living on the street, who attend school rarely, if at all, were not represented in such a sample. To analyze the specific situation and the health behavior and knowledge of this group, a different approach was required. Accordingly, in our second example (see Flick and Röhnsch 2007, 2008), adolescents were selected purposefully at the specific meeting points and hangouts of homeless adolescents and asked for interviews. Participants were aged between 14 and 20 and had no (regular) housing. To gain a more comprehensive understanding of their health knowledge and behavior under the conditions of 'the street', we not only interviewed them but also followed them through phases of their everyday lives, using participant observation. The topics of the interviews were similar to our first example above, with additional questions about the specific situation of living on the street and about how participants entered street life.

In the first part of the study, the sampling and the interviews did not focus on illness. The second part of the study focused on the circumstances of chronically ill homeless adolescents. In addition to interviewing adolescents with various chronic diseases (from asthma to skin diseases and hepatitis), we conducted expert interviews with physicians and social workers in order to obtain their views of the service situation for this target group.

In both of the above examples (Boxes 4.2 and 4.3), the health and social situations of adolescents were studied – either of youth in Germany in general, or of a specific subgroup with particularly stressful conditions of living. In both cases, the results should be useful for helping to prevent health problems in the target groups and to improve the design of services for them.

Characteristics of Research Questions

Research questions may be regarded from different angles. From an external point of view, they should address a socially relevant issue. In our examples, the issues are the health situation of and the support for youth – and in particular the deficits in both.

Are there particularly strong or frequent health problems, and are there service gaps for particular subgroups or for adolescents in general?

Answering the research questions should lead to some kind of progress – through, for example, providing new insights or suggestions for how to solve the problem under study. Thus, documentation of the changes in the health situation across repeated studies can progress the development of knowledge (as in the youth health survey). If you are studying an issue that to date has been analyzed only generally, progress may result from studying the issue with a specific subgroup (as in our second example).

Seen more from an internal point of view, i.e. of social science itself, research questions should be theoretically based, i.e. embedded in a specific research perspective. In the youth health survey, for example, the basis was provided by a model of the links between social structures, the social position of the individuals in those structures, the social and material environments they live in, and behavioral and physiological factors. These links influence the likelihood of suffering from illness and harm, with their respective social consequences (see Richter et al. 2008, p. 14). From this theoretical model, the concrete research questions in the project and then the items in the questionnaire were derived. In the example of the homeless adolescents, the theoretical background was provided by the approach of social representations (see Flick 1998a, 1998b). The core assumption of this approach is that, depending on social context conditions in different social groups, specific forms and contents of knowledge are developed which occur alongside group-specific practices. A further assumption is that topics have specific contents and meanings for each group and its members. These assumptions formed the background for developing the general research questions, which focused on the lived experience of homelessness and the meaning of health within it, and also the specific questions in the interviews.

Research questions should also be suitable for study through the methods of social research. A research question should be formulated in such a way that you can apply one or more of the available methods in answering it – if necessary after adapting or modifying one of them. (We will examine this point further in Chapters 11 and 12.) For example, the research questions of the youth health survey could be investigated by using the questionnaire method. In the second example, the research questions are studied by using two methods, namely (episodic and expert) interviews and participant observation. (These methods will be discussed in detail in Chapter 11.)

Important qualities of research questions are their specificity and their focus. That is, you should formulate your research questions so that they are (a) clear and (b) goal directed, so as to facilitate the exact decisions to be made concerning who or what should be investigated. Note that research questions define not only exactly what to study and how, but also which aspects of an issue may remain excluded. This does not mean that a study cannot pursue several sub-questions; it means only that you should ensure that your research questions are not fuzzy and that your study is not overloaded with too many research questions.

Overall, a number of possible basic research questions in social research can be distinguished, notably (see Flick 2018a, p. 88; Lofland and Lofland 1984):

- What type is it?
- What is its structure?
- How frequent is it?
- What are the causes?
- What are its processes?
- What are its consequences?
- What are people's strategies?

Figure 4.2 arranges these basic research questions in terms of what is happening on the level of practices (processes and strategies), how these practices are embedded (structures and frequencies), what the backgrounds of these practices are (causes and outcomes) and what can be identified as structuring elements (types and patterns).

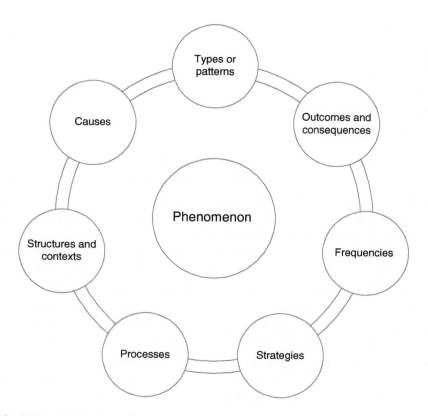

Figure 4.2 Basic research questions

These research questions can be studied at various levels (such as knowledge, practices, situations or institutions) and for different units (for example, persons, groups or communities). Generally speaking, we can differentiate between research questions oriented towards describing states and those describing processes. In the first case, you should describe a certain given state: what types of knowledge about an issue exist in a population? How often can each type of knowledge be identified? In the second case, the aim is to describe how something develops or changes: how has this state come about? Which causes or strategies have led to this state? How is this state maintained – by which structure? What are the causes of such a change? Which processes of development can be observed? What are the consequences of such a change? Which strategies are applied in promoting change?

We can apply these two major types of research question, i.e. those concerning states and those concerning processes, to a variety of study units (see Flick 2018a, pp. 89–90; Lofland and Lofland 1984). For example:

- meanings
- practices
- episodes
- encounters
- roles
- relationships
- groups
- organizations
- lifestyles.

Figure 4.3 arranges these units for focusing research questions in terms of what is happening on the level of activities (practices and meanings), how these activities are embedded (rules, institutions, relations and encounters), what the backgrounds of these practices are (groups, organizations and practices) and what can be identified as structuring elements (differences and commonalities).

We can now begin to differentiate between quantitative and qualitative research in relation to the above lists. The former is more interested in frequencies (and distributions) of phenomena and the reasons for them, whereas qualitative research focuses more on the meanings linked to certain phenomena or on the processes that reveal how people deal with them. In the case of the research projects discussed above, the youth health survey asks for frequencies (item 3 in the first list) and structures (2) in dealing with health in a specific group (14 – here adolescents). The second example focuses on meanings (8) and practices (9) at the level of the participants' strategies (7) with a focus on the types (1) of meanings and the adolescents' practices. All in all, research questions in qualitative research focus more on answering 'why' questions and thus on meanings and

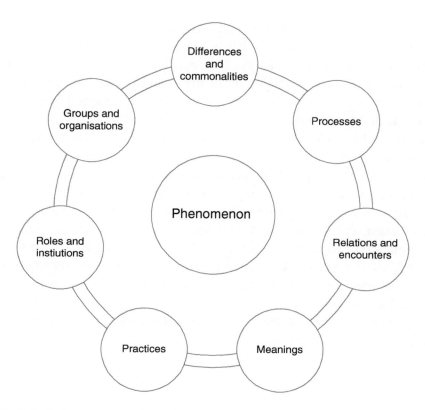

Figure 4.3 Units for focusing research questions

contexts. In quantitative research, they rather focus on answering 'how many', 'how frequent', 'how big' kinds of questions about certain issues.

Good Research Questions, Bad Research Questions

Obviously, you need not only to have a research question, but also to have a good one. Here we will consider what tends to distinguish good research questions from bad ones.

Good research questions

What characterizes a good research question? First of all, it should be an actual *question*. For example, 'The living situation of immigrants from Eastern Europe' is not a research question but an area of interest. 'What characterizes the living situation of immigrants from Eastern Europe?' is a question, but it is too broad and unspecific for orienting a research project. It addresses a variety of subgroups implicitly – and supposes that immigrants from, say, Poland and Russia are in the same situation. Also, the term 'living situation' is too broad; it would be better to focus on a specific aspect of the living situation, such as health problems and the use of professional services. An example might then be: 'What characterizes the health problems and the use of professional services of immigrants from Russia?'

There are three main types of research question: (1) Exploratory questions focus on a given situation or a change, for example: 'Has the health situation of homeless adolescents changed in the last 10 years?'; (2) Descriptive questions aim at a description of a certain situation, state or process, for example: 'Do homeless adolescents come from broken families?' or 'How do adolescents find themselves homeless?'; (3) Explanatory questions focus on a relation. This means that more than just a state of affairs is investigated (so one goes further than asking a question such as 'What characterizes ...?'); rather, a factor or an influence is examined in relation to that situation. For example: 'Is a lack of sufficient specialized health services a major cause of serious medical problems among homeless adolescents?'

Bad research questions

Neuman (2014, p. 175) has characterized what he calls 'bad research questions'. He identifies five types of such questions, which are applied here to our example: (1) questions that cannot be empirically tested or which are non-scientific questions, for example: 'Should adolescents live on the streets?'; (2) statements that include general topics but not a research question, for example: 'Treatment of drug and alcohol abuse of homeless adolescents'; (3) statements that include a set of variables but no questions, for example: 'Homelessness and health'; (4) questions that are too vague or ambitious, for example: 'How can we prevent homelessness among adolescents?'; (5) questions that still need to be more specific, for example: 'Has the health situation of homeless adolescents become worse?'

As these examples may show, it is important to have a research question that really is a question (not a statement) that can be answered. It should be as focused and specific as possible instead of being vague and unspecific. All the elements of a research question should be clearly spelled out instead of remaining broad and full of implicit assumptions. To test your research question before you do your study, reflect on what possible answers to that question would look like.

The Use of Hypotheses

The more explicit and focused your research question is, the easier it is to develop a hypothesis from it. A hypothesis formulates a relation, which will then be tested empirically. (We saw examples of hypotheses in the case study above of Hollingshead and Redlich's 1958 study of social class and mental illness.) Such relations may, for example, be statements taking the form of 'if, then' or 'the more of one, the more of the other'. The first type of relation is found in a hypothesis like: 'If adolescents come from a lower social class, their risk of certain diseases is much higher.' The second form of relation is illustrated by the following hypothesis: 'The lower the social class adolescents come from, the more often they will fall victim to specific diseases.'

Hypotheses should clarify:

- for which area they are valid (are they assumed to be valid always and everywhere or only under some specific local and temporal conditions?)
- to which area of objects or individuals they apply (e.g. humanity as a whole, or just men, or just women younger than 30 years, etc.)
- whether or not they apply to all objects or individuals in this area
- to which issues they apply, i.e. features of the individuals in the object area.

The example of Hollingshead and Redlich's (1958, p. 10) first hypothesis helps to illustrate these features. Their hypothesis was: 'The prevalence of treated mental illness is related significantly to an individual's position in the class structure.' The two issues are 'social class' and 'illness risk'. They have not delimited their hypothesis to specific local or temporal conditions, but have assumed that it should be valid for all the people in a specific social situation (social class).

Furthermore, hypotheses should be clearly formulated in terms of the concepts they use. In our example, what counts as 'mental illness' must be specified (e.g. which diagnoses are used to identify this feature?). What 'treated' means should also be defined (e.g. only people who are being treated with medication, or also those who receive consultations?). Finally, the phrase 'position in the class structure' should be explained.

Hypotheses should also be embedded in a theoretical framework. The authors in our example refer to a number of other studies and theoretical works in which the class structure of America in the 1950s was defined, and they do the same for 'mental illness'.

Hypotheses should be specific, i.e. all predictions included should be made explicit. In our example, the authors did not look at a general relation between illness and social situation; instead they had a specific focus on mental illness and on the position in a hierarchy of five social classes, which they had carefully identified and defined beforehand.

Hypotheses should be formulated in relation to the available methods and have empirical links (how and with which methods can they be tested?). In our example, the hypothesis was formulated such that diagnostic instruments could be used for identifying the illness situation of the participants and such that household surveys could be used for identifying their position in the social class structure.

Just as the research questions that characterize quantitative research may be distinguished from those that characterize qualitative research, so a distinction between the two types of research may be made in terms of the role of hypotheses. Quantitative research should always start from a hypothesis. The procedures in quantitative research are normally oriented towards testing the hypotheses formulated in advance. This means that you should look for empirical pieces of evidence, which allow hypotheses to be either confirmed or contradicted.

Whereas quantitative research starts from hypotheses, they play a minor role in qualitative research. In qualitative research, the aim is not to test a hypothesis that was formulated in advance. In some cases, in the process of research, working hypotheses may be formulated. For example, first observations of differences in reacting to a symptom of skin disease by male and female adolescents living on the street may lead to a working hypothesis that reactions to illness in this context are gendered. Such a working hypothesis will provide an orientation, for which you will search for evidence or counter-examples. But the aim will not be to test this hypothesis in the way you test a hypothesis in a quantitative study. In qualitative research, the use of the word 'hypothesis' has more in common with how you would use this term in everyday life than with the principles of testing hypotheses in quantitative research mentioned above.

Each form of social research should start from a clear research question. Different types of research questions suggest one or the other methodological procedure or may only be answered with specific methods. Hypotheses play various roles: in quantitative research they form an indispensable starting point; in qualitative research sometimes a heuristic tool.

What You Need to Ask Yourself

In planning and doing your empirical project in social research, you should consider the questions in Box 4.4. These questions can be applied to the planning of your own study and, in a similar way, to the evaluation of other researchers' existing studies.

Box 4.4   What you need to ask yourself

Research questions

1   Does your study have a research question which is spelled out clearly?
2   You should be aware of where your research question came from and what you want to achieve with it:

- Is your interest in the contents of the research question the main motivation?
- Or is answering the research question more a means to an end, like obtaining an academic degree?

3   How many research questions does your study have?

- Are there too many?
- Which is the main question?

4   Can your research question be answered?

- What could an answer look like?

5   Can your research question be answered empirically?

- Who can provide insights for that?
- Can you reach these people?
- Where can you find such people?
- Which situations might give you insights for answering your research question, and are these situations accessible?

6   How clearly is your research question formulated?
7   What are the methodological consequences of the research question?

- What resources are needed (e.g. how much time is required)?

8   If necessary for the type of study you chose, did you formulate hypotheses?

- Are they formulated clearly, in a well-defined and testable way?

9   Can your hypotheses be tested? By which methods?

What You Need to Succeed

Box 4.5 lists points that you should have understood and must consider when formulating your research question(s). They should help you to advance with your study successfully.

Box 4.5   What you need to succeed

Research questions

1   Do I understand the importance of a research question?
2   Do I understand the difference between a research question and a research interest?
3   Do I see the difference between a research question and a hypothesis?
4   Do I have an idea now of what distinguishes good from bad research questions?
5   Do I see why a research question should be a question and not a statement?

What you have learned

- Research questions may be developed from practical problems, they may be rooted in the researcher's personal background or they may arise out of social problems.
- Research questions may aim at representative results or at specific subgroups of society.
- Research questions should be embedded theoretically and be ready to be empirically studied. Above all, they should be specific and focused.
- Quantitative research is based on hypotheses which are empirically tested. When qualitative research uses hypotheses, it will be based on a different conception of hypotheses, i.e. as working hypotheses.

What's next

This first text goes into more detail about research questions in quantitative research:

Bryman, A. (2016) *Social Research Methods*, 5th edn. Oxford: Oxford University Press.

This next book deals with linking perspectives in research questions and a number of aspects of how to phrase research questions in greater detail:

Andrews, R. (2003) *Research Questions*. London: Continuum.

The following two chapters discuss research questions for qualitative studies:

Flick, U. (2018) *Designing Qualitative Research*, 2nd edn. London: Sage. Chapter 1.
Flick, U. (2018) *An Introduction to Qualitative Research*, 6th edn. London: Sage. Chapter 6.

This article is based on an empirical study about how research questions within research practice in organization studies are formulated, where such questions come from and which aims drive the selection and formulation of research questions. It also addresses the role of the existing literature in this context, and it provides advice on developing one's own research question:

Sandberg, J. and Alvesson, M. (2010) 'Ways of Constructing Research Questions: Gap-spotting or Problematization?', *Organization*, 18(1): 23–44.

The following case study can be found in the online resources. It should give you an idea of how a relevant problem is identified and turned into a research question for planning and doing a research project:

Scillitoe, J. L. (2019) 'A Personal Journey of Identifying, Developing, and Publishing Survey-based Research in an Emerging Research Area: Incubation of Early Stage Technology Entrepreneurial Ventures', *SAGE Research Methods Cases*. doi: 10.4135/9781526474414.

In this article, also in the online resources, the role and development of research questions are discussed on the basis of empirical knowledge about a target group (here, school nurses intending to do research):

O'Brien, M. J. and DeSisto, M. C. (2013) 'Every Study Begins with a Query: How to Present a Clear Research Question', *NASN School Nurse*, 28(2): 83–5.

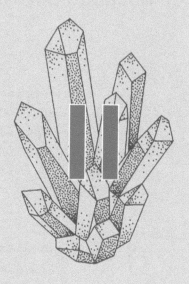

PLANNING AND DESIGN

Part I of this book was designed to help orient you towards doing your research project. Part II guides you through key steps in the early phases of the project itself, whether it be qualitative or quantitative. Chapter 5 considers research literature. It outlines the nature of research literature and how it can – and should – inform the planning of your research project. Chapter 6 offers an overview of the major steps in the research process. This provides the foundations for planning your research project – the subject of Chapter 7. The first practical step in giving your plans a shape is to write a proposal and design a timescale. For this step, it will be helpful to know more about which designs, and which forms of sampling, are used in social research – and the implications of these for your project.

Research Methodology Navigator

Orientation

- Why social research?
- Worldviews in social research
- Ethical issues in social research
- From research idea to research question

You are here in your project

Planning and design

- Reading and reviewing the literature
- Steps in the research process
- Designing social research

Method selection

- Deciding on your methods
- Triangulation and mixed methods

Working with data

- Using existing data
- Collecting new data
- Analyzing data

Reflection and writing

- What is good research? Evaluating your research project
- Writing up research and using results

5

READING AND REVIEWING
THE LITERATURE

You will:

- appreciate the relevance of existing literature for planning your own research project,
- recognize that you should be familiar with the methodological literature as well as with research findings in your area of social research, and
- understand how to find the relevant literature for your research project.

The Scope of a Literature Review

In general, you should begin your research by reading. You should search for, find and read what has been published so far about your issue, about the field of your research and about the methods you wish to apply in your study. (Research methods will be discussed in detail in Chapters 8–12.)

It is also helpful to read in order to understand the basics of social research (discussed in Chapter 1) and the overall process in which it is applied (Chapter 6).

Of course, you cannot read everything that has been said about social research. Fortunately, that is not necessary! However, you should find out what is necessary and helpful for doing a research project on your chosen issue and research question.

Sometimes, you may come across the notion that a qualitative study need not be based on knowledge of the existing theoretical or empirical literature. This notion is, however, based on an outdated conception of what it means to develop a theory from empirical data. Today, there is a consensus amongst qualitative, as well as quantitative, researchers that you should be familiar with the ground on which you move and wish to make progress; finding new insights needs to be based on knowing what is known already.

What Do We Mean by 'Literature'?
Types of literature

When you start researching the issue you have selected, you can search for and find different types of literature and evidence. First, you may find articles in the press. Newspapers and magazines may from time to time take up that issue, perhaps in order to sensationalize it. This kind of literature will help to show what kind of attention the public pays to your issue and perhaps its relevance in public discourse. However, you should be careful not to treat such publications as if they were scientific literature. Figure 5.1 displays the various types of literature you can use.

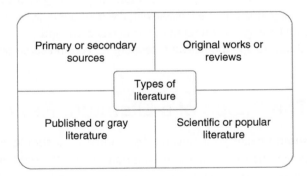

Figure 5.1 Types of literature

Primary and secondary sources

You should take care to distinguish between different types of sources. There are primary sources and secondary sources. An example here may explain this distinction.

Autobiographies are written by the persons themselves. Biographies are written by an author about a person, sometimes without their knowing him or her personally (e.g. if it is a historical person). If we transfer this distinction to scientific literature, a monograph about a theory is a primary source. The same is the case for an article or a book describing the empirical results of a study, when it was written by the researchers who did the study. A textbook summarizing the various theories in a field or giving an overview of the research in a field is a secondary source. Consider too a third example for clarifying this distinction: original documents like death certificates are primary sources, whereas official statistics summarizing causes of death in their frequencies and distribution in groups are secondary sources. The difference rests on how immediate the access is to the reported fact: primary sources are more immediate, whereas in secondary sources usually several primary sources have been summarized, condensed, elaborated on or reworked by others. In general, for central aspects of the study that is planned, it is suggested to go back to the original literature wherever possible and to avoid citations based on other citations. Central aspects refer to other studies in the same area, to the most important theories, to the methods that are employed and to the literature used in discussing the results of one's own study. This means that citations like 'Strauss 1987 cited in Bryman 2016' should be avoided and the original work by Strauss should be quoted – and read beforehand of course.

Original works and reviews

Amongst scientific articles, we can distinguish between articles reporting research results for the first time (e.g. Flick and Röhnsch 2007, which reports on findings from our interviews with homeless adolescents) and review articles (e.g. Kelly and Caputo 2007, which reviews several studies and gives an overview of health and homeless youth in Canada). Similarly, we can distinguish between original publications about a theory and textbooks that summarize theories across a field.

Amongst review articles, we can further distinguish between narrative and systematic reviews. A narrative review gives an account of the literature in the sense of a general overview (as in Kelly and Caputo 2007), including different types of literature (research, government reports, etc.). A systematic review has a stronger focus on research papers, which have been selected according to specific criteria, and has a narrow focus on one aspect of a general issue. An example of a systematic review is the work by Burra et al. (2009) in which a defined range of databases was searched for articles that met a number of criteria predefined in order to assess studies of homeless adults and cognitive functioning. The method of searching (and the choice of databases, criteria, periods of publication, etc.) is specified in order to make the review systematic, replicable and assessable in itself.

An alternative form of summarizing studies is meta-analysis. Again, existing research is reviewed, but here the focus is on the effect of a specific variable and how this can be identified in the studies under analysis. For example, Coldwell and Bender (2007) conducted a meta-analysis of studies about the effectiveness of a specific treatment for homeless people with severe mental illness. Meta-analysis is now also used in qualitative research (see Timulak 2014).

The difference between a literature report and a (systematic) literature review can be outlined as follows: a report concentrates on finding and listing the existing literature in a field of study, while a review is more driven by criteria not only in searching and finding literature, but also in elaborating and presenting it in a structured way. At the end of a literature review, new insights into the research situation have been produced.

Gray literature

In addition to what is published about your specific issue in books and scientific journals, for example, you should look for 'gray' literature, such as practice reports or reflections of practitioners about their work with this target group. These may be reflections in essay form, and sometimes also empirical reports based on numbers of clients, diagnoses, treatment outcomes and the like. Gray literature is defined as 'literature (often of a scientific or technical nature) that is not available through the usual bibliographic sources such as databases or indexes. It can be both in print and, increasingly, electronic formats' (University Library 2009). Examples are technical reports, preprints, working papers, government documents and conference proceedings. This type of literature will often give you more immediate access to ongoing research or debates, as well as to institutional ways of documenting and treating social problems. Thus, gray literature can be a valuable first-hand source for your study.

Finding Literature

In general, it depends on your topic where you should search for and will find relevant literature. If you want to find out whether your usual library holds the literature you are looking for, you can simply go to the library and check the catalogue. This can be time-consuming and frustrating if the book is not in stock. If you want to find out which library holds the book (or journal) you are looking for, you can access the library's online public access catalog (OPAC) via the Internet. Therefore, you should go to the homepage of one or more libraries. Alternatively, you can use a link to several libraries at the same time. Examples are www.copac.ac.uk for over 100 of the major university libraries and the British Library, and https://kvk.bibliothek.kit.edu/index.html?lang=en for most of the German university libraries and also many in the UK and the USA. There you can find an exhaustive overview of the existing books or the information to help complete your reference lists. Many books are now available as e-books, which you can obtain via the library, even from your home or office.

For journal articles, you can use search engines like http://wok.mimas.ac.uk. This will lead you to Thomson Reuters' Social Sciences Citation Index (also accessible through Social Sciences Citation Index) which you can search by author, title, keywords, and so on. Other electronic databases that you can use, if they fit your issue, include PubMed and MEDLINE. PubMed comprises more than 20 million citations for biomedical literature from MEDLINE, life science journals and online books. Citations may include links to full-text content from PubMed Central and publisher websites (www.ncbi.nlm.nih.gov/pubmed). These are databases that document the publications in journals in the field of health and medicine. 'Athens' is an access management system developed by Eduserv that simplifies access to the electronic resources that your organization has subscribed to. Eduserv is a not-for-profit professional IT services group (www.openathens.net). If you want to read a whole article, you may need to buy the right to download it from the publisher of the journal or book. More and more articles are available online and free of charge in open-access repositories (e.g. the Social Science Open Access Repository at www.ssoar.info/); 'open access' means that everybody can use this literature without paying for access.

You may also use online publication services organized by publishing houses such as Sage (the publisher of this book). At https://journals.sagepub.com you can search all the journals published by this publisher, read abstracts and get the exact reference dates free of charge. If you want to read a whole article, you will need to subscribe to the service or the journal, or to buy the article from the homepage (or see whether your library has subscribed to the journal in question).

Of course, a first step in finding your way into the literature can be to use an Internet search engine such as Google, Google Scholar, Intute or AltaVista. This is, however, a first step only and certainly should not be the only one.

In recent years, a lot of effort has been invested in updating the information bases in libraries, along with the instruction on how to use them. This is referred to as promoting research information and data literacies (RIDLs) and can be useful for finding relevant literature in a field coming from various sources (see www.researchinfonet.org for more information on progress in this field).

Areas of Literature

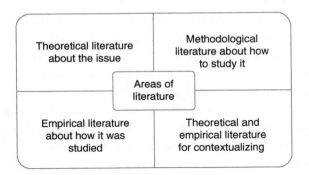

Figure 5.2 Areas of literature

You will need to review the literature in several areas, notably:

- theoretical literature about the topic of your study
- methodological literature about how to do your research and how to use the methods you choose
- empirical literature about previous research in the field of your study or similar fields
- theoretical and empirical literature to help contextualize, compare and generalize your findings.

Figure 5.2 displays the various areas of literature which we will examine one by one.

Reviewing the theoretical literature

Theoretical literature means works on the concepts, definitions and theories used in your field of investigation. Reviewing the theoretical literature in your area of research should help you answer such questions as:

- What is already known about this issue in particular, or about the area in general?
- Which theories are used and discussed in this area?
- What concepts are used or debated?
- What are the theoretical or methodological debates or controversies in this field?

- What questions remain open?
- What has not yet been studied?

Here, you should synthesize the discussion and the concepts and theories that are used in your field of study. The end point should be that it becomes clear which of these have informed your research interest, your study and its design.

We can distinguish several forms of theories. There are those that conceptualize your issue – such as theories of homelessness in our example – and those that define your research perspective, such as social representations (see Flick 1998a). The latter posit that there are different forms of knowledge about an issue linked to different social backgrounds and that these differences are a starting point for analyzing the issue itself.

Reviewing the methodological literature

Before deciding on a specific method for your study, you should read the relevant methodological literature. If you want to use, say, focus groups (discussed in Chapter 11) in a qualitative study, you should obtain a detailed overview of the current state of qualitative research, for example by reading a textbook or an introduction to the field. Next, you should identify the relevant publications on your method of choice by reading a specialized book or by looking at prior examples of research using this method. This will allow you to select your specific method(s) with an appreciation of the existing alternatives. It will also prepare you for the more technical steps of planning to use the method and help you to avoid the pitfalls mentioned in the literature. Such an understanding will help you to compose a detailed and concise account of why and how you used the method in your study, when you write your report later on.

Overall, reviewing the methodological literature in your area of research should help you to answer such questions as:

- What are the methodological traditions, alternatives or controversies here?
- Are there any contradictory ways of using the methods which you could take as a starting point?

Reviewing the empirical literature

In the next step, you should review and summarize the empirical research that has been done in your field of interest. This should allow you to contextualize your approach and, later, your findings and to see both in perspective.

Reviewing the empirical literature in your area of research should help you to answer such questions as:

- What are the methods that are used or debated here?
- Are there any contradictory results and findings which you could take as a starting point?

What You Need to Ask Yourself: Reading Empirical Studies

When you are reading existing studies, it is important to be able to assess them critically – in terms of both their methods and their results. You should consider, in particular, how far the study has achieved its aims, and how far the study meets appropriate methodological standards in collecting and analyzing data. The checklist in Box 5.1 is designed to help you make critical assessments of the literature you read.

Box 5.1 What you need to ask yourself

Reading empirical studies

1 Have the researchers or authors clearly defined the aim and purpose of their study and explained why they conducted it?
2 What is the research question for the study?
3 Did the author review, integrate and summarize the relevant background literature?
4 Which theoretical perspective is the study based on, and has it been explicitly formulated?
5 Which design was applied in the study? Does it fit the research question that was formulated?
6 Which form of sampling was applied? Was it appropriate to the aims and to the research question?
7 Which methods of data collection were applied?
8 How far have ethical issues been taken into account (e.g. informed consent and data protection)?
9 Which methods have been applied for analyzing the data? Is it clear how they were used and maybe modified?
10 Do methods of data collection and analysis fit together?
11 Which approaches and criteria did the researchers apply for assessing their own ways of proceeding?

> 12   Have the results been discussed and classified through reference to earlier studies and the theoretical literature on the research issue?
> 13   Does the study define its area of validity and its limits? Have issues of generalization been addressed?

Kasperiuniene and Zydziunaite (2019) did a systematic literature review of professional identity construction in social media, which can be used as an example of how to find, summarize and analyze the existing literature in a specific field.

How to Construct and Write a Literature Review

Types of argumentation

There are several ways of using the literature you find. First, we should distinguish between, on the one hand, listing literature and, on the other, reviewing or analyzing it. Merely listing what you found and where will not be very helpful. In a review or an analysis of the literature, you will go further than this by ordering the material and producing a critical assessment of it, involving the selection and weighting of the literature.

Structure of a literature review

After searching, finding and reading the literature, you should have an idea of what is relevant for your topic, what the major theories in the field are and which are the most important studies for your own research. You should then develop a thematic and logical structure for the topics your review is to cover. For the example of homeless adolescents' health and chronic illness in Germany (Box 4.3), the literature review could have the following structure (see Box 5.2). The display in the tables in Box 5.2 lists the thematic fields you should cover for a study in this context, with the aim of each of the topics in the second column, and the final column showing the principle this kind of literature review applies (from general to concrete aspects). This structure embeds the concrete focus of the study (homeless adolescents' health) in a more general framework (health research in adolescence) first, before addressing the concrete focus and the methodological approach that will be taken. A crucial step is to identify and name the gaps in the literature, as in section 2.4 in Table 5.2, leading to the research questions to be pursued in your study.

Box 5.2 Research in the real world

Structure of a literature review

In our study on homeless adolescents' health concepts, we did a literature review according to the principle and in the areas listed in the tables in Box 5.2. Table 5.1 focuses on the context of the study.

Table 5.1 Structure of a literature review – the context

Thematic Field	Aim of the Literature Review	Principle
1. Youth and health	Give an overview of the areas in general	From general to concrete
1.1 Health and subjective health concepts	Review the general context of the study (subjective health concepts)	From general context issues
1.2 Selected aspects of adolescents' health	Review the framework in general for adolescents' health: – theoretical concepts and models – earlier studies	
1.3 Managing specific health risks in adolescence	Review the literature for how adolescents deal with health risks: – theoretical concepts and models – earlier studies	
1.4 Managing health problems in adolescence	Review the literature for how adolescents deal with health problems: – theoretical concepts and models – earlier studies	to the
1.5 Chronic diseases in adolescence	Review the literature for which chronic diseases can occur in adolescence: – theoretical concepts and models – earlier studies	
1.6 Living with chronic diseases in adolescence	Review the literature about how adolescents' lives are affected by chronic diseases and how they live with the: – theoretical concepts and models – earlier studies	specific aspects of the context
1.7 Conclusion	Set up the general framework and develop an orientation for the specific focus of the study	

Table 5.2 focuses on the literature about the issue of the study and leads to identifying a gap in the literature and research, and to formulating and the research question(s) for the study.

Table 5.2 Structure of a literature review – the issue

Thematic Field	Aim of the Literature Review	Principle
2. Health and illness in the context of homelessness in adolescence	Give an overview of the specific areas and their links	From general to concrete
2.1 Homeless adolescents' ways of living in Germany and in international comparison	Review the literature for specifying the concrete situation of the target group to be studied in distinction to other contexts: – theoretical concepts and models – earlier studies	From more general issues of the study's topics
2.2 Living on the street: Ways and conditions of living	Review the literature for specifying the concrete situation of living of your target group more generally: – theoretical concepts and models – earlier studies	
2.3 (Adolescent) homelessness and health	Review the literature for specifying the concrete situation of living of your target group in the concrete circumstances you want to study: – theoretical concepts and models – earlier studies	to the
2.4 Conclusions, gaps and research questions of our study	Summarize the main insights from the literature review in the above fields: – identify and state the gaps that have become evident, which justify your study and lead to defining your research question – draw conclusions for the specific target group you want to study – state the main topics of your planned data collection resulting from the literature review	specific aspects of the research question

(Continued)

Table 5.3 focuses on the literature about the methodology to be used in the study, including methodological backgrounds principles and concrete methods.

Table 5.3 Structure of a literature review – the methodology

Thematic Field	Aim of the Literature Review	Principle
3. The study's theoretical and methodological backgrounds	Set your study in context	From general to concrete
3.1 Theoretical framework: Social representations	Outline the theoretical framework as far as necessary for understanding your study: – theoretical concepts and models – earlier studies (examples you use as orientation)	From general methodological issues
3.2 Methodological approaches	Outline the methodological framework as far as necessary for understanding your study: – theoretical concepts and models – methodological principles – earlier studies (examples you use as orientation)	to the specific aspects of the methods to be applied

The outcomes of your analysis may include a synthesis of the range of literature you draw on and some conclusions. These conclusions should lead readers to your own research question and research plan, and provide a rationale for both.

How to build an argument

To illustrate the difference in making a claim and developing an argument, we come back to the research in the real world (see Boxes 4.2 and 4.3) about adolescents' health in the preceding chapter. The building of an argument refers here to the fact that there is a need to study adolescents' health in the context of homelessness. Just to state that this need for research exists is a claim or a big idea. To substantiate this claim as an argument, you should identify evidence for your argument, as in Figure 5.3, from the existing research that you have found, read and analyzed. For example, the existing health surveys covered large samples and addressed issues such as substance use. This would be evidence for the counter-argument – more research about specific subgroups is not necessary. Evidence supporting your argument that more research is necessary could be that the samples did not cover the target group of homeless adolescents and that their particular health risks were not covered explicitly in the questionnaires. Weighing up the evidence pro and con your claim and argument allows for a substantiated statement that additional research is needed and ethically justified (see Chapter 3).

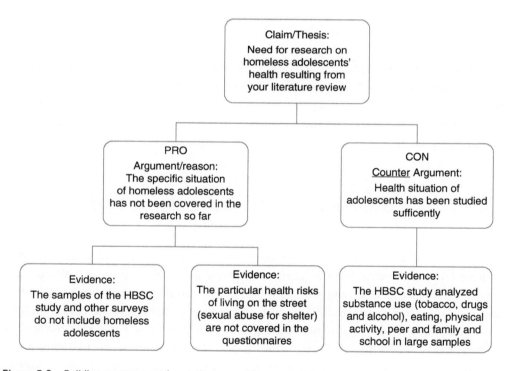

Figure 5.3 Building an argument for stating a need for research

In most cases, it will not be necessary to give a complete account of what has been published in an area. Rather, you should include what is relevant for your project, for justifying and for planning it.

Sometimes it is difficult to decide when to stop working with the literature. One suggestion here is to continue reading and perhaps summarizing new literature while working on the project. You will in any case have to come back to the literature when discussing your own findings. Another suggestion is that you set up your research plan after reviewing the literature and then, from the moment you start your empirical work, try not to be distracted by new literature and not to revise your project continuously.

A concise definition of a literature review's contents is provided by Hart:

> The selection of available documents (both published and unpublished)
> on the topic, which contain information, ideas, data, and evidence written
> from a particular standpoint to fulfil certain aims or express certain views
> on the nature of the topic and how it is to be investigated, and the effective
> evaluation of these documents in relation to the research being proposed.
> (1998, p. 13)

You should demonstrate in the way you present the literature used in your study that you have conducted a skilful search into the existing literature. It should also be evident from your literature review that you have a good command of the subject area and that you understand the issue, the methods you use, and know the state of the art of the research in your field.

Documentation of what you found and read in the literature

It is important to develop a way of documenting what you have read – both the sources and the content. For the latter, you should take notes from the major topics of an article or a book you read and derive from your reading some keywords that you can use for further searches. You should always look at the reference list of what you read as an inspiration for further reading. You can take your notes electronically by typing them up in a file with your word processor, for example, or by using a commercial software tool like Microsoft OneNote (www.office.microsoft.com/en-us/onenote). These tools allow you to store your search results and notes including the sources, i.e. where you found your information. Alternatively, you can make notes manually on index cards. Be sure to note the source of anything you found noteworthy, so that you can come back to the original article and retrieve the context of an argument or concept.

Nothing found and secondary quotations

If your search and analysis of the literature suggest that there is no research or no publications about a topic or an aspect, do not simply state that result, but make it transparent where (in which databases, for example) you have looked for publications.

If you found relevant sources in other publications, which you can not access directly (e.g. because the book or journal are not available in your library), you can use secondary quotations, for example: (James, 2008, as cited in Jones, 2009). Often, only the secondary reference (Jones 2009) will then be listed in the reference list. But you should try to limit the use of secondary quotations as much as possible and to get hold of as many of your sources as you can.

Referencing the Literature

You also need to decide on a system for referencing your literature both in your text and in your reference list.

In-text referencing

Most important here is a systematic and consistent use of literature. You may, for example, take the way I have referred to other sources in this book and the reference list at the end of it as a model for your own use of literature. This means referring to other authors' works *in the text*, as in the following examples:

1 As Allmark (2002) holds …
2 Such an assessment normally considers three aspects: 'scientific quality, the welfare of participants and …' (Allmark 2002, p. 9).
3 Gaiser and Schreiner (2009, p. 14) have listed a number of questions …
4 Flick et al. (2010, p. 755) hold that …

More generally, this means that you refer to other authors' works by using the format of author's name, followed by the year of publication in brackets (as in example 1). When you use a direct quote of the author's (their own words), you will have to add the page number (as in example 2). When you refer to a work by two authors, you mention both authors' names linked with 'and', the year of publication and the page number (as in example 3). If you refer to a work of three or more authors, you mention the first author's name and add 'et al.' and the year and page number (as in example 4).

In the *list of references* at the end of your work, you will mention the material you used as follows:

1   *Book*: Surname and initials of the author(s), year in brackets, book title in italics, place of publication, publisher. Example:

Gaiser, T.J. and Schreiner, A.E. (2009) *A Guide to Conducting Online Research*. London: Sage.

2   *Book chapter*: Surname and initials of the author(s), year in brackets, chapter title, 'in' initials and surname of editor(s), '(ed.)' or '(eds)', book title in italics, place of publication, publisher, first–last page numbers of the chapter. Example:

Harré, R. (1998) 'The Epistemology of Social Representations', in U. Flick (ed.) *Psychology of the Social: Representations in Knowledge and Language*. Cambridge: Cambridge University Press. pp. 129–37.

3   *Journal article*: Surname and initials of the author(s), year in brackets, article title, journal title in italics, volume number, issue number if available, first–last page numbers of the article. Examples:

Allmark, P. (2002) 'The Ethics of Research with Children', *Nurse Researcher*, 10: 7–19.

Flick, U., Garms-Homolová, V. and Röhnsch, G. (2010) '"When they Sleep, they Sleep": Daytime Activities and Sleep Disorders in Nursing Homes', *Journal of Health Psychology*, 15 (5): 755–64.

4   *Internet source*: Surname and initials of the author(s), year in brackets, article title, journal title in italics, volume number, issue number if available, 'Available at:' web link/URL, date on which you accessed the source. Example:

Bampton, R. and Cowton, C.J. (2002) 'The E-Interview', *Forum Qualitative Social Research*, 3 (2), www.qualitative-research.net/fqs/fqs-eng.htm (accessed 22 February 2005).

However, there are four types of referencing formats, which have been widely accepted, although none has been established as *the* referencing format. Helpful overviews can be found at some library sites, for example at Monash University (https://guides.lib.monash.edu/?b=s). Here, some examples in the four most common types of referencing formats are given. APA was developed by the American Psychological Association but is more widely used beyond psychology (https://guides.lib.monash.edu/ld.php?content_id=12586146). The Chicago citation style is used widely for academic writing in the humanities, social sciences and natural sciences (www.chicagomanualofstyle.org/home.

html). Harvard is also widely used in the social sciences (https://guides.lib.monash.edu/ld.php?content_id=8481587) and MLA comes from the Modern Language Association (https://style.mla.org).

First, we will look at in-text citations in the four formats (see Table 5.4).

Table 5.4 In-text citations in APA, Chicago, Harvard and MLA format

Convention	In-Text Citation
APA	Direct quotes: include page or paragraph number, e.g. (Flick, 2005, p. 45) or Flick (2005, p. 45) argues that …
	Two authors: Flick and Röhnsch (2007) argue or: '…' (Flick & Röhnsch, 2007)
	Three to five authors, first time cited: Müller, Meyer, and Schulze (2008) argue or: '…' (Müller, Meyer, and Schulze, 2008) thereafter: Müller et al. (2008) or: '…' (Müller et al., 2008)
	Six or more authors: Only Müller et al. (2008) from the beginning
	Secondary reference, for example: (James, 2008, as cited in Jones, 2009). Only the secondary reference (Jones 2009) will be listed in the reference list
Chicago	(Flick 2005, 45)
	Two authors: '…' (Flick and Röhnsch 2007, 45)
Harvard	'…' (Flick, 2005, p. 45) or Flick (2005, p. 45) argues that …
MLA	See APA

Referencing in the reference list

For journal articles in the reference list see Table 5.5.

Table 5.5 Journal articles in the reference list in APA, Chicago, Harvard and MLA format

Convention	Citation
APA	Prior, L. (2008). Repositioning Documents in Social Research. *Sociology*, *42*(5), 821–836. doi: 10.1177/0038038508094564.
Chicago	Prior, Lindsay. 2008. 'Repositioning Documents in Social Research', *Sociology* 42 (5): 821–836. doi: 10.1177/0038038508094564.
Harvard	Prior, L. (2008). Repositioning Documents in Social Research. *Sociology*, *42*(5), pp. 821–836.
MLA	Prior, Lindsay. 'Repositioning Documents in Social Research', *Sociology*, vol. 42, no. 5, 2008, pp. 821–836. *SAGE Publications*, doi: 10.1177/0038038508094564.

Note that 'doi:' is the 'document online identifier', which every article has now been given. This has become necessary, as more and more articles are published 'online first' before being published in print. The 'doi' for both versions is the same for online and print. The 'doi' is also used for book chapters and books now, as they are also often available online and in print versions.

For book chapters in the reference list see Table 5.6.

Table 5.6 Book chapters in the reference list in APA, Chicago, Harvard and MLA format

Convention	Citation
APA	Bogner, A., Littig, B., & Menz, W. (2018). Generating qualitative data with experts and elites. In U. Flick (ed.), *The SAGE Handbook of Qualitative Data Collection* (pp. 652–665). London: SAGE Publications Ltd. Doi: 10.4135/9781526416070.
Chicago	Bogner, Alexander, Beate Littig and Wolfgang Menz. 'Generating Qualitative Data with Experts and Elites.' In The SAGE Handbook of Qualitative Data Collection, 652–665. London: SAGE Publications Ltd, 2018. Doi: 10.4135/9781526416070.
Harvard	Bogner, A, Littig, B & Menz, W 2018, 'Generating qualitative data with experts and elites', in The Sage Handbook of Qualitative Data Collection, SAGE Publications Ltd, London, pp. 652–665 [Accessed 29 June 2018], doi: 10.4135/9781526416070.
MLA	Bogner, Alexander, et al. 'Generating Qualitative Data with Experts and Elites.' *The SAGE Handbook of Qualitative Data Collection*. Flick, Uwe: SAGE Publications Ltd, 2018, pp. 652–665. SAGE Research Methods. Web. 29 Jun. 2018, doi: 10.4135/9781526416070.

For books in the reference list see Table 5.7.

Table 5.7 Books in the reference list in APA, Chicago, Harvard and MLA format

Convention	Citation
APA	Flick, U. (2018). *An Introduction to Qualitative Research* (6th edn). London: SAGE.
Chicago	Flick, Uwe. *An Introduction to Qualitative Research* (6th edn). London: SAGE, 2018.
Harvard	Flick, U. 2018, *An Introduction to Qualitative Research* (6th edn). SAGE, London.
MLA	Flick, Uwe. *An Introduction to Qualitative Research*. (6th edn). London. 2018.

Alternatively, you may use footnotes for references according to Chicago referencing. Which format you use will depend on preferences – your own and those of your supervisor or faculty. The major point here is that you must work systematically: every journal article must be referenced in the same format and every book cited consistently.

You can also use bibliographic software like EndNote (www.endnote.com) for administering your literature. It will take some time to learn how to use the software, and you should begin using it early in your work and continue to use it while reviewing the literature.

In general, there are two rules to keep in mind:

1   Your use of a referencing style has to be consistent throughout the work you are currently writing. Be consistent in the way you quote and format references.
2   Check with your supervisor and/or institution which style of referencing you should use if it is a thesis or assignment text. If it is an article you are working on, check with the journal (or book editors) which style of referencing is required.

Plagiarism and How to Avoid It

In recent years, the topic of plagiarism has attracted increasing attention in the media and in universities. This is a serious issue, not least because it has become technically much easier to copy and use other people's work.

What is plagiarism?

Plagiarism means that you simply use formulations by other authors without acknowledging them and making it evident that you have quoted them. There are three main forms of plagiarism (see Neville 2010, p. 29): (1) to copy other people's work (i.e. ideas and/or formulations) without quoting the authors; (2) to blend your own arguments with the ideas and words of other people without referring to them; and (3) to paraphrase other authors' formulations without referring to them, and pretending it was your own work.

Inadvertent plagiarism

Plagiarism can occur for several reasons. The most obvious one is that people intend to use someone else's ideas and/or formulations and to pretend that they were their own ideas or formulations. In this case, they would be aware of their plagiarism. This can also be seen as an intentional deceit. However, there may be other reasons, such as that someone does not know what plagiarism is, or is insecure about how to quote correctly, or is careless in the use of other people's materials. This is referred to as 'inadvertent plagiarism', which in the end will have the same consequences as intentional plagiarism.

Why you should avoid plagiarism

In general, plagiarism offends the rules of good practice in scientific work and it is illegal. Plagiarism detection software is more frequently used now for identifying the use of other authors' formulations without quoting the authors explicitly. Once it is detected, the student or researcher will face very serious consequences, such as failing in their thesis or being removed from the university.

How to avoid plagiarism

There are several ways to avoid plagiarism (including inadvertent plagiarism). The first is to pay sufficient attention to including a full list of all the references you have used in writing your thesis. The second is to be very thorough in quoting when you use other people's words. Thus, you should put all other authors' wordings you use in quotation marks. You should also use suspension points (ellipses) when you leave certain words out of a quote, and you should indicate when you add a word by putting it in square brackets: for example, 'social research … [is] valuable'. If you take a sentence from another author and paraphrase it, so that the same content and ideas are still the basis of your formulations, and you do not mention the original author and source, this still constitutes plagiarism. If you take a quotation from a text in which another text was already quoted, you should notify this secondary quotation: for example, '(Author 2, as cited in Author 1, pp. 182–93)'. Both should go in the references.

To avoid plagiarism, you should take your own thoughts and formulations as the basis of your thesis, document your sources carefully and keep notes of where you have found and read something. Finally, you should use more than one source for developing your arguments.

What You Need to Ask Yourself: Using Literature

In composing a literature review for an empirical project in social research, you should consider the questions provided in the checklist in Box 5.3.

Box 5.3 What you need to ask yourself

Using literature

1. Is your literature review up to date?
2. Is your literature review connected to your research issue?
3. Is your literature review and your writing about it systematic?
4. Does it cover the most important theories, concepts and definitions?
5. Is it based on the most relevant studies in your field of research and about your issue?
6. Did you document how and where you searched for literature?
7. Do your research question and design result from your review of the literature?
8. Are they consistent with it?
9. Have you handled quotations and sources carefully?
10. Did you summarize or synthesize the literature you found?
11. Did you take care to avoid plagiarism?

What You Need to Succeed

The questions in Box 5.4 should help you to see if you are ready for your study in what concerns the use of literature.

Box 5.4 What you need to succeed

Using literature

1. Do I understand the relevance of doing a literature review?
2. Do I see the different areas of relevant literature?
3. Did I develop an argument from my literature review which leads to my research question?
4. Does this argument provide justification for doing my research in addition to the existing body of research?
5. If there is no literature about my topic, did I document how I searched for publications?

What you have learned

- In social research, the search and analysis of existing literature are the most important steps.
- There are several points in the research process where the use of literature can be helpful and necessary.
- In planning research, in analyzing materials and in writing about findings, you should make use of existing literature about (a) other research, (b) theories, and (c) the methods you use in your study.

What's next

These first two books provide a comprehensive overview of how to do a literature review for your study, which pitfalls to avoid and how to write about what you find:

Hart, C. (1998) *Doing a Literature Review*. London: Sage.
Hart, C. (2001) *Doing a Literature Search*. London: Sage.

This third text explains how to use references and avoid plagiarism:

Neville, C. (2010) *Complete Guide to Referencing and Avoiding Plagiarism*. Maidenhead: Open University Press.

The next book provides information about forms of gray literature and how to use it:

Fink, A. (2018) *Conducting Research Literature Reviews: From the Internet to Paper* (5th edn). London: Sage.

The following case study can be found in the online resources. It should give you an idea of how to do a literature review for a research project in a more systematic way:

Postăvaru, G. and Cramer, D. (2016) 'A Case of Methodological Premises underlying Literature Reviews', *SAGE Research Methods Cases*. doi: 10.4135/9781446273050155595385.

The following article, also in the online resources is an example of a systematic literature review:

Kasperiuniene, J. and Zydziunaite, V. (2019) 'A Systematic Literature Review on Professional Identity Construction in Social Media', *SAGE Open*. https://doi.org/10.1177/2158244019828847.

Research Methodology Navigator

Orientation

- Why social research?
- Worldviews in social research
- Ethical issues in social research
- From research idea to research question

You are here in your project

Planning and design

- Reading and reviewing the literature
- Steps in the research process
- Designing social research

Method selection

- Deciding on your methods
- Triangulation and mixed methods

Working with data

- Using existing data
- Collecting new data
- Analyzing data

Reflection and writing

- What is good research? Evaluating your research project
- Writing up research and using results

6

STEPS IN THE RESEARCH PROCESS

—How this chapter will help you—

You will:

- obtain an overview of the process of social research
- appreciate, from the point of view of planning, what qualitative and quantitative research have in common and also how they differ, and
- develop an understanding of which steps in the research process you need to take into account when planning your project.

Overview of the Research Process

The differences in worldviews in the background of social research discussed in Chapter 2 lead to differences in planning concrete research projects. Oriented on critical rationalism, in quantitative research, the research process is planned primarily in a linear way: one step follows another in sequence. Oriented on social constructionism, in qualitative research, the process is less linear: some of these steps are more closely interlinked, while some are omitted or located at a different stage of the process. This chapter provides an outline of the research process for both approaches.

The Research Process in Quantitative Research

Some of the processes outlined in this chapter are rather abstract. To understand them, it will help to have a concrete case. Imagine as an example that staff at a hospital have noticed that, in its everyday routines, waiting times are too long. To analyze when and under what conditions delays occur, and to find a possible solution to the problem, the institution can go two ways: either it commissions a systematic study of situations in which waiting times are produced by a researcher from outside (e.g. a sociologist or psychologist); or someone on the team is commissioned to undertake this task. If your area is not health, imagine a different institution (say a school) with a similar problem (say waiting times in appointments for parents with teachers) to illustrate the research steps outlined below.

Step 1: Selection of a research problem

Every research project begins with the identification and selection of a research problem. We have seen in Chapter 4 that the potential sources of research problems are diverse. In our example here, the motivation for the research results from a practical problem.

Step 2: Systematic searching of the literature

As the next step, a systematic review of the literature is required. This should cover three areas, namely:

- theories about the issue or theoretically relevant literature (see Chapter 5)
- other studies about this or similar issues
- overviews of the relevant literature on research methods in general or on specific methods (see Chapter 8).

In our example, the researchers should look for theoretical models concerning how waiting times emerge and concerning routines in institutions of this kind. They should also focus on empirical studies about the organization of processes in hospitals and comparable service enterprises (e.g. schools).

Step 3: Formulation of a research question

It is not enough to identify a research problem. For virtually all research problems, one could study a number of research questions – but not all at the same time. Thus, the next steps are: to decide on a specific research question; to formulate it in detail; and, above all, to narrow it down.

In our example, the research question could be either 'When do waiting times occur particularly often?' or 'What characterizes situations in which waiting times occur?' Depending on the research question, qualitative or quantitative research or a specific qualitative or quantitative method will be (more) appropriate (see Chapter 4).

Step 4: Formulation of a hypothesis

In quantitative research based on critical rationalism, the next step, after formulating the research question, is to formulate a hypothesis. In our example, a hypothesis could be: 'Waiting times occur more often after the weekend compared to other days in the week.' A hypothesis formulates a relation in a way that can be tested (see Chapters 2 and 4).

Step 5: Operationalization

To test a hypothesis, you first have to operationalize it. This means that you transform it into entities that can be measured or observed or into questions that can be answered. In our example, you should operationalize the term 'waiting time' and define more concretely how to measure it. How much time (e.g. more than 10 minutes) is seen as waiting time? When should one start to measure waiting time? At the same time, for our hypothesis, we should define what is meant by 'after the weekend' (e.g. the time from 9.00 a.m. to 12.00 p.m. on Mondays).

Step 6: Development of a project plan or research design

The next step is to develop a project plan and a research design for your study. First, you choose one of the usual designs, and then define how to standardize and control the processes in your study so that you can interpret any relations you find in an unambiguous way. In our example, you could test our hypothesis first by comparing waiting times at the beginning of the week with those on days like Thursday and Friday (see Chapter 7).

Step 7: Application of a sampling procedure

To make this research design and project plan work, you need to define which groups, cases, or fields should be integrated into your study. Such decisions in selecting empirical units are made by applying sampling procedures. In our example, you could take a period of four weeks and include all wards of a hospital on Mondays and Thursdays. To define in which wards you collect the data on both days, you can use a randomized procedure (see Chapter 7).

Step 8: Selection of the appropriate methods

Next, you need to select appropriate methods for collecting and analyzing data. Here you have three basic alternatives:

- First, you can select one of the existing methods. For example, you may use a questionnaire, which has been applied successfully by other researchers.
- Often, you will have to modify an existing method. For example, you may delete some questions in that questionnaire and/or add new ones.
- If that is not sufficient for the study you are planning, the third option is to develop your own method, such as a new questionnaire, or a new methodology, such as a new form of interviewing (see Chapter 11).

In our example, observations in the selected wards would provide time measures of how long patients wait after admission to the hospital for the beginning of assessment or treatment on each ward. At the same time, you may interview team members on the ward using a questionnaire, addressing how often such waiting times occur according to their experience and what they see as the reasons (see Chapter 9).

Step 9: Access to the research site

Once you have finished the planning of your study on the methodological level, the next step – especially if you are conducting applied research – is to find a site in which you can do the study. Here, normally four tasks have to be solved. First, if your research is expected to take place in an institution, you have to organize access to the institution as a whole. Second, you should gain access to individuals in the institution for participation. Third, you need to clarify issues of permission. Fourth, questions of how to protect participants from any misuse of the data and of anonymity have to be addressed.

Consider again our example. You have first to find and select appropriate hospitals (i.e. that are relevant to the research question). In each hospital, the director's (and perhaps the staff council's) agreement is required before you can approach particular wards and their managers. Once a hospital has agreed to join the study, you must persuade individual members of the institution to participate in the study – according to the sample that is intended (for example, your sample should include not only trainees but also experienced doctors and nurses, maybe even in leading positions). With respect to data protection, you should clarify how to prevent the identification of particular participants within the institution and of the institution as a whole from the outside. For example, you could develop a system of nicknames or a different form of anonymization. Finally, it should be made clear who is allowed to see the data and in what form, and who is not allowed. Similar procedures would be necessary if you intend to do your research in a school.

Step 10: Data collection

After finishing these methodological preparations, you are ready to start collecting your data. Basically, you can choose between three main options (see Chapter 11). You may do a survey in one form or another (e.g. with a questionnaire). You may carry out observations (in our example, with a measurement of waiting times). Or you may analyze existing documents (in our example, documentation of treatment routines could be analyzed for waiting times, which can be found in or reconstructed from those documents – see Chapter 10).

Step 11: Documentation of data

Before you can analyze your data, you have to decide how to document them. A survey can be filled in by participants or, alternatively, completed by the researcher based on the respondents' answers. The form of documentation will have an influence on the contents and the quality of the data. Therefore, it should be identical for all participants.

The next step is to edit the data. Take the example of questionnaires. First, the completed questionnaires themselves need to be entered onto your database. Questionnaires often include questions with open answers where respondents write down their answers in their own words. These answers must then be coded. This means they have to be summarized into a number of types of answers and each of these types has to be allocated a number. This elaboration of the data has an influence on their quality and should be done in an identical way for each case (each questionnaire, each observation). This makes the standardization of data and procedures possible (see Chapter 11).

Step 12: Analysis of data

Data analysis constitutes a major step in any project. In quantitative studies, the coding of data is essential. This requires filing responses into categories defined in advance or allocated numerical values (like choosing a response between 1 and 5 in Figure 1.1). If you cannot define the categories in advance, the data (statements, observations) now have to be categorized. This means that identical or similar statements are summarized in a category. This leads to the development of a category system (see Chapter 12 for more details).

In a quantitative study, coding and categorization are followed by statistical analysis. In our example, you could refer the average extension of the measured waiting times to the day of the week when they were measured, or differentiate them for the wards in which they were measured. Responses in questionnaires are, for instance, analyzed for their frequencies and distribution in different professional groups.

Step 13: Interpretation of results

Not every statistical analysis produces meaningful results. How you interpret any relationships that you find in the data therefore becomes very important. Note that when a statistical analysis shows that certain events occur together or that they correlate in their frequencies or intensities, this does not provide any explanation for why this is the case and what that means in concrete terms. If you show, in our example, that waiting times on Mondays are much longer than those on Thursdays, this result does not provide any explanation as to why this is so. If estimations of waiting times differ between nurses and physicians, for example, we need to find an explanation for this finding (see Chapter 12).

Step 14: Discussion of the findings and their interpretations

The analysis and interpretation of data and results are followed by a discussion of them. This means that findings are linked to existing literature on the issue (or the methodology used) and to other relevant studies. Could our study find anything new? Has it confirmed what was known already or did contradictions of existing results emerge? What does it mean if the hypothesis that waiting times occur more often or are extended after the weekend could not be verified, and if this result is different from what was found in other studies? How do the staff's perceptions of the problem found here differ from the results of similar surveys by other researchers? In the context of such a discussion, you will continue to look for explanations for what you found in your data.

Step 15: Evaluation and generalization

From a methodological point of view, this is particularly relevant. You will critically assess your results and the methods that led to them, which means checking their reliability and validity (see Chapter 13 for more details). At the same time, you should check what kind of generalization the results justify, asking: can they be transferred to other fields, to other samples, to a specific population (e.g. all nurses, all wards, all hospitals; or all hospitals, physicians, etc. of a specific type)? A major element here is to make transparent what the limits of the results are, such as what could not be found or confirmed, or what limits are set for transferability to other institutions (see Chapter 13).

Step 16: Presentation of the results and the study

Whether the results of a study are recognized depends mainly on the way in which the results are presented. This means first summarizing the study and its main results and then writing (a report, an article, a book, etc.) about it. Often, it is also necessary to select the important information (and to leave out what is less important) due to limits of space and of the reading capacities of potential audiences. In presenting both results and the process, it is also important that you manage to make the process of the study transparent to readers and thus allow them to assess this process and its results (see Chapter 14).

Step 17: Use of the results

In applied fields of research – such as health or education – the question arises of how to use the results. This means formulating implications or recommendations. For instance, how can you develop suggestions for changing routines on the ward from analyzing the waiting times in the everyday life of the ward? Using the results can also mean applying them in practical contexts – for example, where a specific theoretical model (used in the study) is applied in some wards. Finally, 'use' can mean testing the results against the conditions in practical work and thus evaluating them critically.

Step 18: Development of new research questions and the identification of a new study

A major outcome of an empirical study is the formulation of new research questions or hypotheses for analyzing the questions that remained unanswered, so that research in the field can progress. The new research questions may then lead to the identification of a new study.

Summary

These steps of the research process in quantitative research can be seen as a linear process as they usually follow one after the other. In practice, however, the process is often more recursive: you will find you often need to go back and revise a step when it becomes evident that an earlier decision does not work in practice. Table 6.1 summarizes the steps of the research process in quantitative research. You can use this outline of the research process in quantitative research for planning your empirical study. It can also be helpful as a framework for reading and assessing existing empirical studies.

Table 6.1 Steps of the process in quantitative research

1	Selection of a research problem
2	Systematic searching of the literature
3	Formulation of the research question
4	Formulation of a hypothesis
5	Operationalization
6	Development of a project plan or research design
7	Application of a sampling procedure
8	Selection of the appropriate methods
9	Access to the research site
10	Data collection or use of existing data
11	Documentation of data
12	Analysis of data
13	Interpretation of results
14	Discussion of the findings and their interpretations
15	Evaluation and generalization
16	Presentation of the results and the study
17	Use of the results
18	Development of new research questions and the identification of a new study

---

Box 6.1   Research in the real world

Students' research on an educational intervention for reducing anti-Semitic attitudes in schools

In their master's thesis research in education, Elizaveta Firsova and Marco Schmidt (2017) reviewed the international literature on evaluations of interventions based on Allport's (1954) contact hypothesis approach in order to reduce anti-Semitism. They then evaluated a state program for reducing prejudice in Berlin schools with eight treatment group classes of N = 139 students and five comparison classes without treatment including N = 106 students. They used a standardized assessment instrument and variance analysis of repeated measures to show a reduction in prejudice. Despite an increase in tolerance and knowledge about inter-religious topics, no significant change in attitudes could be found.

The Research Process in Qualitative Research

As we have seen, the process of conducting quantitative research may be broken down into a linear sequence of conceptual, methodological and empirical steps. The particular steps can be presented and applied one after the other and are more or less independent of each other. In qualitative research, these phases may also be relevant, but they are more closely connected. These connections are illustrated in the following discussion; the step numbers are specified in Table 6.2. Again, we can take an example to illustrate this process. Imagine that some specific social groups do not receive the appropriate healthcare support they need. To understand why this lack of support occurs and which specific circumstances play a role in this context, a group of sociologists is commissioned to undertake a research project.

Step 1: Selection of a research problem

First, you will have to select a research problem for such a qualitative study. In our case, this could be the healthcare of homeless adolescents in a large German city and why it is not adequately provided.

Step 2: Systematic searching of the literature

Although the importance of being familiar with the existing literature and research is sometimes downplayed, for such a study it is very important to analyze the existing literature in the three areas mentioned above: theories, other studies, the methodological literature. In our example, the researchers should look for other studies describing the health situation of homeless adolescents (or their situation in general), such as in other countries, and look at literature describing the origins and processes of becoming homeless.

Step 3: Formulation of the research question

Although the research problem mentioned as an example here is rather specific already, it is necessary to decide on a specific research question and to formulate it in a detailed and focused way.

In our example, the research question could be 'What are the barriers preventing homeless adolescents from gaining access to existing healthcare services?' To answer this question, qualitative research or a specific qualitative method (e.g. interviews) might be appropriate. If so, formulating a hypothesis will not be part of the research process, as qualitative research is not aimed at testing hypotheses formulated beforehand. Nor is operationalization a part of the planning of qualitative research, although researchers in our example would think about which questions to ask in their interviews in order to understand what the barriers might consist of, which questions might reveal relevant experiences of the adolescents, and the like.

Step 4: Development of a project plan or research design

Again, this next step is different for a qualitative study. Here, the aim is not to standardize and control the processes in the study but to plan which groups to interview and compare, where to find them and whether interviewing the adolescents is enough to understand the complex social problem under study here.

Step 5: Selection of the appropriate methods

Thus, a major issue in a qualitative project is to select the appropriate methods for understanding the issue under study. The guiding principle here is that the method should be open enough to discover the perspectives of the field or its members, and to allow detailed descriptions of the processes within it. Again, you can select an existing method (a type of interview; in rare cases, even the questions asked in an earlier study), modify an existing method (take the type of interview and its principles but formulate new questions, maybe new kinds of questions) or develop a new method (see Chapter 11).

In our example, we could use an existing method (e.g. the episodic interview – see Chapter 11) but modify it by formulating specific questions about the experience of health services.

Step 6: Access to the research site

In qualitative studies, as in our example, we are often confronted with the problem of how to access 'hard-to-reach' people (such as homeless adolescents) in very specific, sometimes not very accessible, contexts (the hangout spots of adolescents without a regular place to stay, etc.). Here, the process of finding entry into such a field is often very instructive in understanding that field, the social relations within it and the process of integrating new members. In this process of access, it often becomes clear which methods are viable in the context to be studied and how those methods should be adapted.

In our example, it may be necessary to use gatekeepers (e.g. staff of services the adolescents might be in contact with) to gain access to and locate the adolescents. It may also be helpful to start with an extended period of participation (in an ethnographic participant observation kind of way) to build up relations and familiarity with members in the field before approaching individual members for an interview.

Step 7: Sampling of cases

Sampling in qualitative research is much more focused on concrete cases than in quantitative research, where cases are selected in more formal ways (e.g. in a randomized way – see Chapter 7). In qualitative research, the individuals to be studied are selected according to their relevance to the research topic. They are not (randomly) selected to construct a (statistically) representative sample of a general population. In our example, specific cases, i.e. adolescents in a specific situation, are selected and approached directly for an interview. Then other cases are selected, again for specific features.

Step 8: Data collection

Again, we can pursue three major routes to collecting data in qualitative research. You may ask people (in interviews), you can carry out (participant or non-participant) observations in the field and you can use existing documents for understanding, in our example, how the relations between adolescents and services work, and which perspectives and practices are involved (see also Chapters 10 and 11).

Step 9: Documentation of data

In the documentation of data in qualitative research, recording is normally the first step. Interviews, for example, are recorded on tape, mp3 or video. In observations, field notes or protocols are written up, sometimes based on video recordings. For interviews, transcription, i.e. making a written text of what was recorded orally, is the next step (see Chapter 11).

Step 10: Analysis of data

In qualitative research, data analysis mainly involves the interpretation of interview statements, events or actions documented in field notes made from observations. Here, you will also look for explanations – why some statements occur in specific contexts together with other statements, or why they occur more often in certain circumstances (see Chapter 12).

Steps 7–10: Sampling, collection, documentation and analysis of data

In qualitative research, the central parts of the research process tend to be more interrelated than in quantitative research. They can also be planned and organized differently compared to quantitative research.

A more strongly interlinked version of this process has been developed by Glaser and Strauss (1967) in their grounded theory approach (see also Strauss 1987, Strauss and Corbin 1990 and Flick 2018e). The use of grounded theory within qualitative research makes the latter very different from quantitative research. The aim is to do empirical research in order to use the data and their analysis to develop a theory of the issue under study. Thus, theory is not a starting point for research, but rather the intended outcome of the study.

This has consequences for the planning of and for the steps taken in the research process. The grounded theory approach gives priority to the data and the field under study over theoretical assumptions. Theories are less applicable to the subject being studied.

Rather, they are 'discovered' and formulated in working with the field and the empirical data to be found in it. Individuals to be studied are selected according to their relevance to the research topic. They are not (randomly) selected to construct a (statistically) representative sample of a general population. The aim is not to reduce complexity by breaking it down into variables, but rather to increase complexity by including context. Methods also have to be appropriate to the issue under study and have to be chosen accordingly. This approach focuses strongly on the interpretation of data no matter how they were collected. Decisions on the data to be integrated and the methods to be used for this are based on the state of the developing theory after analyzing the data already at hand at that moment (see Chapter 12).

In our example, researchers would begin with rather unstructured observations in one spot and talk to members (service providers' staff, homeless adolescents, etc.) more or less informally. After analyzing their first observation and interview data, they would select and include another spot and continue observations and conversations there and analyze the resulting materials. They would continue to include further cases and make comparisons among all cases so far included.

Step 11: Discussion of the findings and their interpretations

As with quantitative research, collection and analysis of the data (see Chapters 11 and 12) will lead to a discussion of the results and their quality (see Chapter 13). Qualitative results should also be discussed in the light of the existing (or lacking) research prior to the study, for their novelty and limitations. In our example, we would try to demonstrate the common aspects and differences with studies in other countries or referring to older homeless people.

Step 12: Evaluation and generalization

The discussion about appropriate criteria for evaluating qualitative studies is still ongoing (see Chapter 13 for more on this). The main issues here are how far the methods and their applications to the field under study are appropriate, and how far they lead to new and meaningful insights. Generalization in qualitative research is not based on statistical representativeness but pursues the question of how far results from one field can be transferred to other fields – for example, how far a theory developed from studying the problems of service utilization by homeless adolescents can be transferred to other vulnerable groups.

Step 13: Presentation and use of results and new research questions

In many qualitative studies, one major aim is to produce relevant results that can be used (implemented) in the fields that have been studied. In our example, such an implementation would aim at reducing the identified barriers to service utilization for homeless adolescents. Thus, presenting and using the results and developing new research questions will be the final steps here, too.

Comparing the Processes of Quantitative and Qualitative Research

Qualitative research is compatible with the traditional, linear logic of quantitative empirical research only to an extent. Interlinking the empirical steps according to the model of Glaser and Strauss (1967) tends to be more appropriate to the character of qualitative research, which is more oriented towards exploring and discovering what is new. Figure 6.1 summarizes the differences between the two approaches. In quantitative research, you have a linear (step-by-step) process, starting from theory and ending with a validation of the theory based on testing it. Sampling is usually finished before data collection begins – which means you have fixed your sampling frame before you send out a questionnaire, for example. It only makes sense to begin interpreting the data – for example, a statistical analysis – once the data collection has been done (see Chapters 11 and 12). Therefore, these steps can be seen as forming a sequence and you will work through them one after the other, as represented by the linear model in the upper half of Figure 6.1.

In qualitative research, these steps are sometimes interlinked. Sampling decisions are taken during data collection, and interpretation of the data should begin immediately with the first data – for example, the first interview. From analyzing these data, you may arrive at new decisions, such as whom to interview next (thus, in this example, the process of sampling continues). You will also immediately start to compare your data – for example, the second with the first interview, and so on. The aim here is to develop a theory from the empirical material and analysis, where the starting point was preliminary assumptions about the issue you wanted to study. This process is represented by the circular model in the lower part of Figure 6.1.

Despite this interlinking of some essential steps in the process of the research, you can also see the qualitative research process to some extent as forming a sequence of decisions (see Flick 2018a). These decisions refer to selecting the specific method(s) of data collection and analysis to be applied, the sampling procedure used in the concrete study (see below) and the ways of documenting and presenting the results.

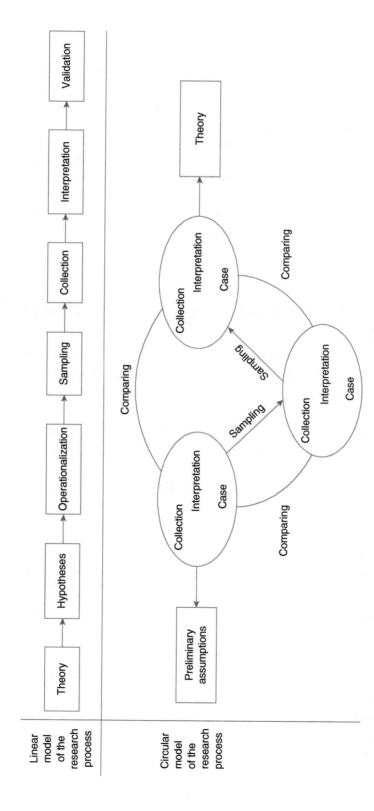

Figure 6.1 Process models of quantitative and qualitative research

Source: Flick 2018a, p. 129

Of course, the two approaches can be combined, for example in a triangulation or mixed methods study (see Chapter 9). However, to make such a combination fruitful and to avoid stumbling blocks coming from confusing the underlying logic, it is important to understand the differences between the two process models.

Table 6.2 juxtaposes the concepts of the research process in quantitative and qualitative research, as outlined in this chapter, in a comparative way. This indicates the main differences in the research process between qualitative and quantitative research. These mainly involve the degree of standardization of procedures. This table also provides a basis for planning the steps of your own project and for developing an adequate design for your study (see Chapter 7 for more detail on this).

Table 6.2 Steps of the process in quantitative and qualitative research

Quantitative research	Qualitative research
1 Selection of a research problem	1 Selection of a research problem
2 Systematic searching of the literature	2 Systematic searching of the literature
3 Formulation of the research question	3 Formulation of the research question
4 Formulation of a hypothesis	4 Development of a project plan or research design
5 Operationalization	5 Selection of the appropriate methods
6 Development of a project plan or research design	6 Access to the research site
7 Application of a sampling procedure	7 Sampling of cases
8 Selection of the appropriate methods	8 Data collection or use of existing data
9 Access to the research site	9 Documentation of data
10 Data collection or use of existing data	10 Analysis of data
11 Documentation of data	Or: Sampling, data collection, documentation of data, analysis of data, comparison of data, sampling of data, data collection, documentation of data, analysis of data, comparison, etc.
12 Analysis of data	
13 Interpretation of results	
14 Discussion of the findings and their interpretations	11 Discussion of the findings and their interpretations
15 Evaluation and generalization	12 Evaluation and generalization
16 Presentation of the results and the study	13 Presentation and use of results and new research questions
17 Use of the results	
18 Development of new research questions and the identification of a new study	

What You Need to Ask Yourself

The questions in Box 6.2 should help you in planning your research.

Box 6.2 What you need to ask yourself

Planning research

1 How far did you reflect on the decisions in the research project outlined in this chapter?
2 Did you take the differences between quantitative and qualitative research into account?
3 How consistent are the decisions you make in your research?
4 Are these decisions appropriate for your overall research question?
5 Do they take into account the ethical dimensions of your research?

What You Need to Succeed

In planning your own empirical project, you should take into account the aspects shown in Box 6.3 and answer the questions that arise. This checklist may help you in planning your own study, but you can also use it to assess the existing studies of other researchers.

Box 6.3 What you need to succeed

Planning research

1 Be aware of which steps in the research process are appropriate for the kind of study you are planning.
2 Seek to establish whether the existing knowledge and research about your issue are sufficient for planning and doing a quantitative study in which you will test hypotheses.
3 Or seek to clarify whether the empirical and theoretical knowledge about your study topic is so limited or has so many gaps that it makes sense to plan and do a qualitative study.
4 Check the procedures in your plan for their soundness. Does the methodological plan of your study fit (a) the aims of your study, (b) its theoretical background, and (c) the state of the research?
5 Will this kind of research process be compatible with the issue you want to study and with the field in which you intend to do your research?

What you have learned

- Quantitative and qualitative studies run through some similar and some different steps in the research process.
- Quantitative research is planned in a linear process.
- In qualitative research, in many cases the steps in the research process will be interlinked.
- Having a firm grasp of the steps that comprise the research process in each case will give you a basis for planning your own study.

What's next

Overviews of planning social research in a quantitative approach are discussed in the first and fourth books here, while the middle two readings focus on this issue in the context of qualitative studies:

Bryman, A. (2016) *Social Research Methods*, 5th edn. Oxford: Oxford University Press.
Flick, U. (2018) *Designing Qualitative Research*. London: Sage.
Flick, U. (2018) *An Introduction to Qualitative Research*, 6th edn. London: Sage. Part 3.
Neuman, W.L. (2014) *Social Research Methods: Qualitative and Quantitative Approaches*, 7th edn. Essex: Pearson.

This short book provides a comparative introduction to the different versions of grounded theory research and the process model of qualitative research discussed in this chapter:

Flick, U. (2018) *Doing Grounded Theory*. London: Sage.

This next article describes the qualitative research process from an ethical and epistemological point of view:

Sörensson, E. and Kalman, H. (2018) 'Care and Concern in the Research Process: Meeting Ethical and Epistemological Challenges through Multiple Engagements and Dialogue with Research Subjects', *Qualitative Research*, 18(6): 706–21.

The following case study can be found in the online resources, it gives you an idea of why it is important to test your instruments in qualitative studies:

Mikuska, E. (2017) 'The Importance of Piloting or Pre-testing Semi-structured Interviews and Narratives', *SAGE Research Methods Cases*. doi: 10.4135/9781473977754.

The following article can also be found in the online resources and gives you an idea of why and how a relevant problem is turned into a quantitative study:

Stephens, C., Noone, J. and Alpass, F. (2014) 'Upstream and Downstream Correlates of Older People's Engagement in Social Networks: What are their Effects on Health over Time?', *International Journal of Aging and Human Development*, 78(2): 149–69.

Research Methodology Navigator

Orientation

- Why social research?
- Worldviews in social research
- Ethical issues in social research
- From research idea to research question

You are here in your project

Planning and design

- Reading and reviewing the literature
- Steps in the research process
- Designing social research

Method selection

- Deciding on your methods
- Triangulation and mixed methods

Working with data

- Using existing data
- Collecting new data
- Analyzing data

Reflection and writing

- What is good research? Evaluating your research project
- Writing up research and using results

7

DESIGNING SOCIAL RESEARCH

┤How this chapter will help you├

You will:

- understand how and why to write a proposal and develop a timescale
- obtain an overview of the most important research designs
- understand the procedures for selecting study participants, and
- appreciate the special role of designs in qualitative research.

Chapter 6 provided an outline of the research process. We can now use this to develop a foundation for research planning.

Writing a Proposal for a Research Project

Planning social research will become concrete when preparing your final thesis in a university program. In most cases, you will need a proposal to enrol for your thesis as well as to apply for a grant. If this is not required, writing a proposal can nevertheless be an important and helpful step in planning your project and in estimating whether it is realistic under the given conditions, such as the time and the skills you have. A proposal for an empirical project should include topics and sub-items as follows (see Table 7.1).

In the introduction, you should briefly outline the background of the project (why you intend to do it) and the relevance of the topic. In the description of the research problem, you should summarize the state of the art in research and in the literature and derive your own research interest from the gaps that become evident in this summary. What emphasis you should put on each of the points will depend on the type of study you plan. The purpose of the study and the aim of doing it should be briefly described. A major point in any proposal is the research question (see Chapter 4), if possible and necessary, divided into major questions and sub-questions.

Whether or not it is necessary to formulate a hypothesis will depend on the type of research you intend to do. For a quantitative study in the context of a critical rationalist worldview, this step should always be included as testing a hypothesis is the central aim of empirical projects. In any case, you need to describe the methodological procedures you intend to apply. For qualitative research, you should give a short justification of why you will use qualitative methods and why you will work with a specific method. In quantitative research, such a justification is often not necessary although it is helpful for reflecting on the plan for your project. The research strategy – an exploratory, a hypotheses testing or an evaluative approach – should be described as well. In the proposal, you should outline the research design (see below) in terms of its main features: which sample will be included, how big will it be, and on which comparative perspective is it based? In the next step, your proposal should include a short description of and justification for the methods you intend to use for collecting your data (see Chapters 8 and 11), for analyzing your data (see Chapters 8 and 12) and for assessing the quality of your study (see Chapter 13). In many contexts, it is a requirement that a proposal covers ethical issues (data protection, non-maleficence, informed consent, etc. – see Chapter 3) and demonstrates how the researchers intend to take them into account in carrying out their project.

In this context, you should consider what results you expect from your study and what will be their relevance in the light of earlier results and practical issues (see Chapters 1 and 14). You will elaborate on the state of research in writing your thesis or the final report of your research in more detail. Nevertheless, I would strongly recommend that you get a first overview of the literature while planning your project, in order to ensure that you don't adopt a research question which has already been answered before to a sufficient extent. Finally, you should briefly discuss the practical conditions of doing your study. For this purpose, it will be helpful to develop a timescale (see below) and to outline your own experience with research. At the end, a preliminary list of references should be added.

Spelling out the topics listed in Table 7.1 is helpful for developing your research project (and the thesis based upon it) and should make successful work more likely.

Abdulai and Owusu-Ansah (2014) developed a systematic account of what to include in a research proposal, which can be taken as an orientation.

Table 7.1   Model for a proposal structure

1   Introduction
2   Research problem

   (a)   Existing literature
   (b)   Gaps in the existing research
   (c)   Research interest

3   Purpose of the study
4   Research questions
5   Methods and procedures

   (a)   Characteristics of qualitative research and why it is appropriate here
   (b)   Research strategy
   (c)   Research design

      (i)   Sampling
      (ii)   Comparison
      (iii)   Expected number of participants, cases, sites, documents

   (d)   Methods of data collection
   (e)   Methods of data analysis
   (f)   Quality issues

6   Ethical issues
7   Expected results
8   Significance, relevance, practical implications of the study
9   Preliminary pilot findings, earlier research, experience of the researcher(s)
10   Your own experience with doing social research
11   Timeline, proposed budget
12   References

Developing a Timescale

A timescale for your project should outline both the steps required in the research process (see Chapter 6) and the time estimated for each step. You may also indicate milestones, which means the outcomes that are to be expected once each step is completed. In Box 7.1, you will find an example of a timescale for a qualitative study using interviews and participant observation (it comes from our study on homeless adolescents' health concepts – see Flick and Röhnsch 2007). Such a timescale can have two functions. In a proposal forming an application for funding, it will demonstrate how much time is needed and for what, in order to convince the funding agency that the budget you are asking for is justified. In designing the research (see below), the timescale will help to orient you towards planning the project.

Box 7.1 Research in the real world

Timescale for a research project

Work Step	Month of Project																							
	1	2	3	4	5	6	7	8	9	10	11	12	13	14	15	16	17	18	19	20	21	22	23	24
Literature research	■	■	■	■	🕮																			
Development of instruments and pre-test				■	■																			
Fieldwork: Finding participants and data collection (e.g. interviews)							■	■	■															
Transcription										■	■													
Fieldwork: Participant observation							■	■	■	■														
Writing observation protocols					■	■	■	■	■	■	■													
Analysis of interviews												■	■	■	■	■	🕮							
Analysis of observation protocols											■						■	■						
Linking back the results to the literature																			■	■	■			
Final report and publications																					■	■	■	🕮

🕮 = Milestone

Source: Flick and Röhnsch 2007

Box 7.2 Research in the real world

Timescale for a student project

Work Step	Week of Project																							
	1	2	3	4	5	6	7	8	9	10	11	12	13	14	15	16	17	18	19	20	21	22	23	24
Literature research	■	■	■	■	⊞																			
Development of instruments and pre-test				■	■	■																		
Fieldwork: Finding participants and data collection (e.g. interviews)						■	■	■	■															
Transcription									■	■	■													
Analysis of interviews										■	■	■	■	■	■	■	⊞							
Linking back the results to the literature																	■	■	■	■	■			
Writing the thesis																	■	■	■	■	■	■	■	⊞

⊞ = Milestone

The timescale in Box 7.1 gives an orientation for a 'professional' project with external funding. For a student project, a two-year plan may be too extended, as students often have a short time frame for research projects for a master's thesis, for example. More realistic will be six months. For this purpose, an adapted suggestion is made next, in Box 7.2.

To make the research (and, beforehand, the proposal) work, the following guidelines should be kept in mind:

- You should try to make the design of your research and the methods as explicit, clear and detailed as possible.
- The research questions, and the relevance of planned procedures and expected data and results for answering them, should also be as explicit and clear as possible.
- The study and the expected results and implications should be placed in their academic and practical contexts.
- Ethics and procedures should be reflected as far as possible.
- Methods should be made explicit not only in the how (of their use) but also in the why (of their selection).
- Ensure that plans, timelines, existing experiences and competences, methods and resources all fit into a sound program for your research.

Designing a Study

A central concept in research planning is the research design, which Ragin defines as follows:

> Research design is a plan for collecting and analyzing evidence that will make it possible for the investigator to answer whatever questions he or she has posed. The design of an investigation touches almost all aspects of the research, from the minute details of data collection to the selection of the techniques of data analysis. (1994, p. 191)

Blaikie and Priest (2019 p. 36) suggest a framework for designing research. A research design will have to answer three basic questions: What will be studied? Why will it be studied? How will it be studied? For the last question, they specify five further questions: What is the logic of inquiry? What are the epistemological assumptions that are adopted? Where will the data come from? How will the data be collected and analyzed? And when will each stage of the research be carried out? Blaikie and Priest have developed four research design tasks – focusing the purpose of the study (e.g. the problem to be studied, the questions to be answered), framing (logics and epistemologies), selecting (data types, forms and sources) and distilling (the timing and doing of collecting and analyzing data and answering the questions).

Research Designs in Quantitative Research

When you construct or use a specific research design in quantitative research, the aims are first to make answering the research question possible and to control the procedures. 'Control' here refers to the means of keeping the conditions of the study constant, so that differences in the replies of two participants can be rooted in their own differences (in their attitudes, for example) and do not result from the fact that they were asked in different ways. This in turn requires you to keep the conditions of the study constant and to define your sampling procedures (who is selected and why – see below).

Research designs in quantitative research are often constructed around analyzing the relations between specific variables. The term 'variable' simply refers 'to an attribute on which cases vary' (Bryman 2016, p. 42); for example, people have different ages or incomes. Age and income can be variable, and different ages can be related to the number of diseases people in a study suffer from. Variables are things that influence something else: first is the independent variable and second is the dependent variable, as it is understood as being dependent on the influence of the first. Variables can, furthermore, be conceived as confounding variables (Bryman 2016, p. 345), or as external variables which may influence the relation of two variables that are being studied. Research designs are often constructed to control the influence of such external variables – to rule out the notion that they are biasing the results.

Quantitative research in a critical rationalist perspective is aimed at producing data that can be seen as facts independent of the research situation and the researcher. Here, another aim of research designs is to control external variables. This refers to factors which are not part of the relations that are studied but influence the phenomenon under study. If you study the effects of a medication on the course of a certain disease, you should make sure that other factors (e.g. specific features of some patients or eating during the treatment) do not influence the course of the disease. One way to control such external variables is to use homogeneous samples (see below). Thus, you will select patients who are very similar in their most relevant characteristics (e.g. men of 50–55 years in specific professions). The disadvantage of such a homogeneous sample is that you can only generalize the results to people who also fulfill the criteria of the sample (men of 50–55 years in specific professions). It is more consistent to draw a random sample (see below). The disadvantage here is that the sample will need to be rather sizeable if specific features (of subgroups of people) are to be represented in it.

A second way of controlling external influences is to use consistent methods of data collection. This means that the data are collected from all participants in the same way in order to guarantee that differences in results come from differences in participants' attitudes and not from differences in the data collection situation. Some of the most common designs in empirical research are presented below.

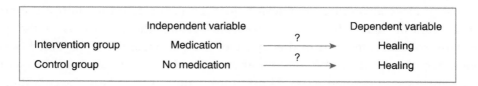

Figure 7.1   Control group design

Control group designs

Let us again begin with an example. In the practice of a hospital, it might be observed that for patients who received a certain medication, the symptoms are reduced and have disappeared by the end of the treatment. This indicates that healing has occurred. To discover whether the improvement in the patients' state is caused by the medication under study, control group designs are often applied.

In such a design, the medication is labeled the independent variable. The improvement in the patients' state that it produces is the dependent variable. Note that the term 'dependent' variable refers to the variable which is caused or changed by another variable (the independent variable). In our example, the healing depends on giving the medication. Thus, the medication is labeled independent, as it is not influenced by the other variable in this setting; the healing has no effect on the medication.

To find out whether such a relation really exists – that the healing depends on the medication (and not on something else) – you will apply a control group design. Two groups of patients are selected; the members of these groups are comparable in features like diagnoses, age and gender. The intervention group receives the medication under study. The control group is not given the medication or receives a placebo (a pill without substance and effects). At the end of treating the first group, you can compare the two groups to see whether the first group has improved over the control group. If in the end both groups show the same changes, the effect cannot be attributed to the medication (see Figure 7.1).

Experimental designs

Control group designs also form the basis of experimental research. Here, to continue our example, the aim is less to find out the effects of a medication than to test the effects of taking it. Experiments consist of goal-directed acts upon study groups in order to analyze the effects of these acts. An experimental design includes at least two experimental groups, to which participants are randomly allocated. The independent variable is manipulated by the researcher (Diekmann 2007, p. 337).

In an experimental study, in contrast to a control group design, the medication (independent variable) is given or changed for research purposes. In order to be sure that an observed effect (of healing, for example – the dependent variable) does not come from the patient knowing 'I am being treated', participants are not informed in such a study as to whether they are given the medication or a placebo without effect. Because the patient does not know whether he has received medication or a placebo, this design is also called a blind test.

If, in our example, a clear difference becomes evident – that the healing effects in the intervention group can be documented for significantly more patients than in the control group – this could still be due to another influence. It might not be the medication in itself that produces the healing effect, but rather the attention paid to participants by nurses or doctors associated with the administration of the pill. To be able to exclude this influence, in many medication trials a double-blind test is applied. Here, both groups are given treatment, with the intervention group receiving the medication and the control group a placebo, which look identical. To avoid any influence by the staff giving the medication – for example, any subconscious signals sending the message that the placebo is without effect anyway – you would not inform the doctors and nurses which is the pill and which the placebo. Neither patients nor staff know who has received the medication and who was given the placebo. Thus, this design is called a double-blind test.

To exclude the influence of other variables as far as possible, experiments are often conducted in laboratories. They are run not in the service routines or everyday life of the hospital, but in an artificially designed setting. This allows control or exclusion of any external influence as far as possible. The disadvantage is that the results are difficult to transfer to contexts outside of the laboratory – i.e. to everyday life.

Pre–post design

Another way of excluding external influencing variables is to do a pre–post measurement with both groups. First, the initial situation is measured in both groups (pre-test measurement). Then the study group is given the treatment (intervention). Afterwards, for both groups, the post-test situation is documented. A problem here is that the first measurement might influence the second one, for example due to a learning effect. This means that experience with the tests, measurements or questions might facilitate the reaction in the second measurement – for instance, a question can be answered more easily (see Figure 7.2).

| Pre-test | Intervention | Post-test |

Figure 7.2 Pre–post design

Cross-sectional and longitudinal studies

The designs presented so far aim at controlling the conditions of the study. This allows us to capture the state of things at the moment of the study. Most studies are planned to take a picture of the moment: interviews or surveys, say, are done at one moment in time, for example to analyze the attitude of a specific group (the French) towards a specific object (a political party). Such picturing is based on a cross-sectional design – a measurement is done to capture the state at a specific moment. In most cases, you take a comparative perspective, such as by comparing the attitudes of several subgroups – for example, those who voted for one political party and those who voted for another party in an election.

However, if processes, courses or developments form the focus of the study, such a cross-sectional design will not be sufficient. Instead, you should plan a longitudinal study to document a development – for example, the attitudes of one or more groups over the years. This attitude is measured repeatedly – for instance, the same instruments are used every two years with the same samples in order to find out how attitudes towards a specific political party have changed. If you repeat such surveys not just once but several times, you can produce time series or trend analyses to document long-term changes in political attitudes.

Longitudinal studies are also interesting if you want to study the influence of a specific event on attitudes or the life course. An example is how the mental illness of a family member develops and how it influences other family members' attitudes to mental illness in general over the years. A problem in this context could be that this process of changing attitudes can only be covered in a comprehensive way when attitudes have been measured for the first time before the illness occurred. If this is not possible, often a retrospective study is done instead – after the illness has been diagnosed, family members are asked about their attitude to mental illness both before the illness of the family member and now.

Qualitative Research Designs

Qualitative research pays less attention to research designs and even less to controlling conditions by constructing specific designs. It aims much more to create an environment in which the views of participants or the making of social situations can be analyzed and understood. This has to do with the worldviews 'social constructionism' and the 'interpretative paradigm', discussed in Chapter 2. In general, in qualitative research, the use of the term 'research design' refers to the planning of a study: how to plan data collection and analysis and how to select empirical 'material' (situations, cases, individuals, etc.) in order to be able to answer the research question in the available time and with the available resources.

The literature on research designs in qualitative research (see also Flick 2004a, 2018b or 2018a, Chapter 7) addresses the issue from two angles. Creswell (1998) presents a number of basic models of qualitative research from which researchers can select one for their concrete study. Maxwell (2005) discusses the parts of which a research design is constructed. (See Flick 2018a, Chapter 7 for more details of what follows.)

Case studies

The aim of case studies is precise description or reconstruction of cases (for more detail, see Ragin and Becker 1992 and Stake 1995). The term 'case' is understood rather broadly here. You can take persons, social communities (e.g. families), organizations or institutions (e.g. a nursing home) as the subject of a case analysis. Stake (1995, pp. 3–4) suggests a differentiation of three types of case studies in the context of school research, for example: 'intrinsic case studies' try to develop an empirical understanding of a particular teacher – for example, their teaching practices. The focus remains on this particular case and on its internal structure (it is therefore 'intrinsic'). 'Instrumental case studies' use the single case to understand some bigger issue beyond the particular teacher that is studied empirically, such as how marking students in school affects the teachers' teaching (p. 3). The third type Stake suggests are 'collective case studies', where, for example, several teachers from one school are studied to understand the practices (or the climate) in that school.

Your main problem then will be to identify a case that would be significant for your research question and to clarify what else belongs to the case and what methodological approaches its reconstruction requires. If your case study is concerned with the learning problems of a child, you have to clarify, for instance, whether or not it is enough to observe the child in the school environment. Do you need to integrate an observation of the family and its everyday life? Is it necessary to interview teachers and/or fellow pupils? Cases in general can be defined on several levels – as individuals (the child), as social entities (the family) or as institutions (the school), for example. Cases then are studied for their uniqueness and their commonalities (Stake 1995, p. 1).

Comparative studies

Often, rather than observe some single case as a whole and in all its complexity, you will instead observe a multiplicity of cases, focusing on particular aspects. For example, you might compare the specific content of the expert knowledge of a number of people in respect of a concrete experience of illness. Or you might compare biographies of people with a specific illness and their subsequent course of life. Here arises the question of the selection of cases in the groups to be compared.

A further issue is the degree of standardization or constancy required for those conditions that you are not focusing on. For example, in order to show cultural differences in views of health among Portuguese and German women in one study conducted by Beate Hoose and Petra Sitta (see Flick et al. 1998), interview partners from both cultures were selected. This ensured that, in as many respects as possible, they lived in at least very similar conditions (e.g. comparable professions, income and level of education) in order to be able to relate differences to the comparative dimension of 'culture' (see Flick 2000b).

Box 7.3   Research in the real world

A student's research study on the relevance of spirituality in hospice work

For her master's thesis in education, Franziska Wächter (2019) studied the relevance of a topic for professional work in health care. Her focus was on spirituality (religious beliefs, for example) for working in end-of-life care in a hospice. Her research questions were: (1) Which meanings of spirituality can be reconstructed from the views of palliative carers (social workers and nurses) in hospices? (2) What is the role of the professional background in this? Wächter did episodic interviews (see Chapter 11) with five nurses and five social workers. Her sampling was via institutions and complemented by snowballing (see below). Her interviews covered five areas: questions and narratives about (1) the perception of spirituality in hospices, (2) subjective understandings of spirituality, (3) forms and ways of spiritual care, (4) spiritual care and interprofessional collaboration, and (5) perceptions and evaluations of spirituality in hospice work. In comparing the two professional groups of nurses and social workers, she was able to identify differences in the representations and evaluations of spiritual components in working in end-of-life care, which could be linked to the professions.

Retrospective studies

Case reconstruction is characteristic of a great number of biographical investigations that examine a series of case analyses in a comparative, typology-oriented or contrastive manner (see Chapter 12). Biographical research is an example of a retrospective research design in which, retrospectively from the point in time when the research is carried out, certain events and processes are analyzed in respect of their meaning for individual or collective life histories. Design questions in relation to retrospective research involve the selection of informants who will be meaningful for the process to be investigated. They also involve defining appropriate groups for comparison, justifying the boundaries of the time to be investigated, checking the research question and deciding which (historical) sources and documents (see Chapter 10) should be used in addition to interviews. Another issue is how to consider the influence of present views on the perception and evaluation of earlier experiences.

Snapshots: analysis of state and process at the time of the investigation

Qualitative research will often focus on snapshots. For example, you might collect different manifestations of the expertise that exists in a particular field at the time of the research in interviews and compare them to one another. Even if certain examples from earlier periods of time affect the interviews, your research does not aim primarily at the retrospective reconstruction of a process. It is concerned rather with giving a description of circumstances at the time of the research.

Longitudinal studies

Qualitative research may involve longitudinal studies, in which one returns to analyze a process or state again at later times of data collection. Interviews are conducted repeatedly and observations are extended, sometimes over a very lengthy period.

Figure 7.3 presents the basic designs of qualitative research.

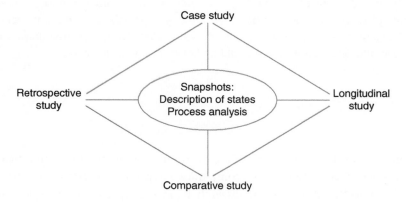

Figure 7.3   Basic designs in qualitative research

Source: Flick 2018a, p. 117

In what follows, I will examine one aspect of planning empirical studies in more detail, namely the question of how to select participants for a study so that the insights produced will be (generally) valid and representative. Either you start with a big group (e.g. German youth), which cannot be studied empirically in its entirety, and select from this group your cases for study so that results can later be generalized to the original group or population (this addresses the issue of statistical sampling in quantitative research), or you select those cases that are particularly relevant for answering your research question (this addresses the issue of purposive sampling in qualitative research).

Sampling

Most empirical studies involve making a selection from a group for which propositions will be advanced at the end. If you want to study the professional stress of nurses, you will need to address a selection from all the nurses in the UK, for example, as it is too big a group to be studied in its entirety. 'All nurses in the UK' is the basic population from which you will draw a sample for your study: 'The population is the mass of individuals, cases, events to which the statements of the study will refer and which has to be delimited unambiguously beforehand with regards to the research question and the operationalization' (Kromrey 2006, p. 269).

In exceptional cases, you can use the strategy of complete collection, which means all cases of a population are included in the study. The other extreme is to select and study one person in a single case study (see above). In most studies, a sample will be drawn according to one or other of the procedures described below and the results are then generalized to the population. Arguments against complete collection and for sampling are that the latter saves you time and money and allows for greater accuracy.

Here, you should distinguish between sampling elements and empirical units. The latter refers to the units which you include in your data collection. For example, the sampling elements are several hospitals in which you want to study the emergence of waiting times. The empirical units are specific situations in these hospitals in which waiting times occur or are to be expected, e.g. situations of preparing for surgery. The problem of sampling arises in qualitative and in quantitative studies in a similar way, but it is addressed differently: statistical sampling is typical of quantitative research, while qualitative researchers apply procedures of purposive or theoretical sampling.

Sampling Strategies in Quantitative Research

There are a number of requirements for a sample. The sample should be a minimized representation of the population in terms of the heterogeneity of the elements and the representativeness of the variables. The elements of the sample have to be defined. The population should be clear and empirically defined. This means that the population has to be clearly limited. Here, we find two alternatives for sampling: random and non-random procedures (see Figure 7.4).

Simple random samples

Often, you do not know enough about the constitution or features of the population to make a purposive selection such that the sample is a minimized representation of the population. In such cases, it is suggested to draw a random sample. Here, we can distinguish between simple and complex random sampling. An example of simple random sampling is selecting from a card index. The elements of a population are documented in a list or card index; for example, all inhabitants of a town are registered in a card index at the residents' registration office. You can use this card index for drawing a simple or systematic random sample. A simple random sample results when every

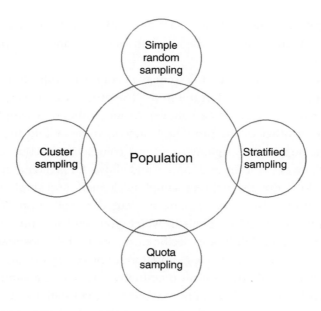

Figure 7.4 Sampling strategies in quantitative research

element in the sample is drawn independently in a random process from the population. One example here is to use a lottery drum: all postcards in a competition are in a lottery drum and are drawn one after the other. Each time, the cards are mixed again before the next is drawn. If we transfer this principle to drawing a sample from a residents' registration file, you would give a number to every entry in the file and make a ticket for each number. All the tickets are then mixed in a lottery drum. One after the other you would draw numbers until your sample is complete. This process can be simulated with a computer.

For very large populations, systematic sampling is suggested. The first choice (the first case, the first index card) is selected randomly (by having a throw or by randomly picking a number from a random number table). The other elements to be included in the sample are defined systematically. For example, from a population of 100,000, you draw a sample of 1000 elements. The first number is selected randomly between 1 and 100, e.g. 37. Then, you include systematically every 100th case in the sample (i.e. the cases 137, 237, 337, ..., 937).

One disadvantage of such a simple random sample is the difficulty in representing relevant small subpopulations in the sample. In such sampling, you will also neglect the context of the particular case, i.e. its features beyond the main criterion for its selection, and will lose that for the analyses. To give an example for the first disadvantage – in a population of nurses from which you want to draw a sample, you find an ethnic minority which could be of particular interest for your study. In simple random sampling, the chances are very limited that members of that subgroup will be included at all or in sufficient number. Context information (concerning, say, ethnic minorities) cannot be taken into account systematically in a simple random sample without neglecting its principle.

Systematic random sampling: stratified and cluster sampling

Therefore, more complex or systematic forms of random sampling may be applied by drawing a stratified sample. The intention is often to be able to analyze the data in the sample separately for specific groups – e.g. to compare the data of the ethnic minority with those of all participants. Accordingly, you will divide the population into several subpopulations (in our example, according to the members' ethnic background). From each of these subpopulations, you will then draw a (in most cases simple) random sample. In our example, you would divide the population of nurses according to their ethnic backgrounds and then draw a random sample from the subgroups of the British, Turkish, African, Korean, etc. nurses. If you apply the same sampling procedure to every group and if you take into account the proportion of each subgroup in the population, you will receive a proportionally stratified sample. In the sample, the percentage of each subgroup is exactly the same as in the population. In our example: if you know that the proportion of Turkish nurses is 20% of all nurses and that of the Koreans is 5%, you will draw random samples for each of these groups in the population until you have 20% Turkish and 5% Korean nurses in your sample.

In small samples, the consequence here is that the real number of cases with a Korean background will be very small – for example, one case in a sample size of $n = 20$. This is not sufficient for statistical analyses comparing Korean and other nurses. As a solution, you could extend the sample so that in every subsample there will be enough cases, which will increase the financial and time resources necessary for the study. An alternative is to build a disproportionate stratified sample; this would balance the under-representation of a subsample. The sample is drawn in such a way that for every subsample the same number of cases is included. In our example, you would aim at a sample size of 20 cases and in each subgroup you would apply random sampling until you have five cases for each group (British, Turkish, African, Korean).

To take contexts more strongly into account than in simple random samples, cluster sampling can be applied. In research in schools, you will select students not as empirical elements but in subgroups like school classes, in which you will collect data and make statements for every member (the individual students). The students are the empirical elements (to whom you apply a questionnaire) and the classes are the sampling elements. You will only talk of a cluster sample when the empirical units are not the clusters themselves (i.e. the classes) but their particular members (the pupils).

You can also draw samples in stages, working on a number of levels. For example, you will first make an overview of the location of all schools for nursing, from which you draw a simple random sample. Then you will divide the selected schools into units of roughly the same size, such as classes. From all of these classes, you draw another random sample. This sample is then divided into subgroups according to their performance, for example (all students with an average grade of better than 5, all with grades between 4 and 2, and all with 2 and worse). From these subgroups, you will again draw a random sample, which finally constitutes the group of those given a questionnaire.

The big advantage of random sampling is that samples drawn in this way are representative of all features of the empirical elements. A non-random sample can only claim to be representative of the features according to which it was drawn.

Non-random sampling: haphazard, purposive and quota sampling

It is not always possible or even desirable to draw a random sample. Nevertheless, sampling should be as systematic as possible. A relatively unsystematic method of sampling is haphazard sampling. Here, we do not have a defined sampling plan or framework, according to which we decide which elements of the population are integrated into the sample. An example is the person-on-the-street interview, in which everyone passing by at a specific moment and ready to be interviewed is integrated into the sample. The decision is taken haphazardly, i.e. not according to defined criteria.

A different strategy is purposive sampling. For example, you carry out a study in which experts will be interviewed, and define criteria according to which someone is an expert on your study topic (or not). You will then search for individuals who meet these criteria. If their number is big enough, you can apply a questionnaire to them, which will be analyzed statistically. The sampling from the population of all experts for this issue is not random. Due to the applied criteria, this is not a haphazard sampling either. In most cases, however, you have to assume that the experts are typical cases (of experts). The problem then arises of how to decide whether the individual case is a typical case or not. Often, this definition is set by the researchers. For this definition, they need enough knowledge about the population to be able to decide whether a case is typical or not. Whether experts are typical experts for this issue can often only be decided at the end of the study by comparing them with other experts. Therefore, the sampling here is often done by using substitute criteria, such as professional experience in a specific position. But this again assumes a link between expertise and professional experience.

In addition, you may apply the concentration principle. This means that in the sampling you focus on those cases which are particularly important for the study topic. The cases are selected according to their relevance – either the very rare cases, which have the strongest influence on the process under study, or those cases that can be found most often, for example.

Survey research often uses the technique of quota sampling. Here, specific features (e.g. age and gender) are defined by which the participants should be characterized. For these features, you will then define quotas of the values of these features, which will be represented in the sample at the end. For example, the distribution of gender might be four to six, i.e. in 10 interviews four men and six women will be included. Age groups might be distributed in such a way that in 10 interviews two participants are younger than 30 years, three are older than 60, and five are between 30 and 60. In these quotas, you will then seek out participants haphazardly – i.e. not randomly. If this procedure works, some conditions should be specified. The distribution of the quota features (in our example, age and gender) in the population has to be known. A sufficient relation between the quota features and the features to be studied (e.g. health behavior) has to be given or assumed. The quota features have to be easy to assess.

Finally, you can use the principle of snowballing – that is, you ask your way from the first participant to the next ('Who else do you think could be appropriate for this study?'). Often, for practical reasons, this is the best or only way to arrive at a sample. The representivity of this sample, however, is rather limited.

Sampling Strategies in Qualitative Research

In qualitative research, some of the sampling strategies discussed above are applied. Others, however, such as random sampling, are seldom found. Here again you can use quotas (of age or gender) or haphazard or purposive sampling (of experts, for example). Principles of snowballing or of concentration are also used. Some unique principles can refine and systematize these approaches for the specific aims of qualitative research (see Figure 7.5).

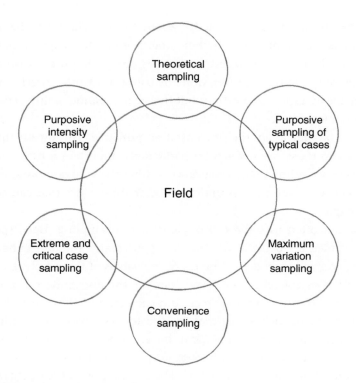

Figure 7.5 Sampling strategies in qualitative research

Theoretical sampling

If the aim of the research is to develop a theory, sampling strategies are likely to be based on 'theoretical sampling', as developed by Glaser and Strauss (1967). Decisions about choosing and putting together empirical material (cases, groups, institutions, etc.) are made in the process of collecting and interpreting data. Glaser and Strauss describe this strategy as follows:

> Theoretical sampling is the process of data collection for generating theory whereby the analyst jointly collects, codes and analyzes his data and decides what data to collect next and where to find them, in order to develop his theory as it emerges. This process of data collection is controlled by the emerging theory. (1967, p. 45)

Here, you select individuals, groups, and so on according to their (expected) level of new insights for the developing theory in relation to the state of theory elaboration so far.

Sampling decisions aim at the material that promises the greatest insights, viewed in the light of material already used and the knowledge drawn from it. The main question for selecting data is: '*What* groups or subgroups does one turn to *next* in data collection? And for *what* theoretical purpose? ... The possibilities of multiple comparisons are infinite, and so groups must be chosen according to theoretical criteria' (Glaser and Strauss 1967, p. 47).

Sampling and the integration of further material are complete when the 'theoretical saturation' of a category or group of cases has been reached (i.e. nothing new emerges any more). In contrast to a statistically oriented sampling, theoretical sampling does not refer to a population whose extent and features are already known. You also cannot define in advance how big the sample to be studied has to be. The features of both the population and the sample can only be defined at the end of the empirical study on the basis of the theory that was developed within it. Wiedemann (1995, p. 441) has compared statistical and theoretical sampling for some essential differences. As his comparison shows, in statistical sampling much more is defined in advance than in theoretical sampling. In theoretical sampling, researchers will know what characterizes the population (its extensions, its features and their distribution in it) they are studying only after the sampling – and thus the research – has been completed. In statistical sampling, the extension and distribution of features in the basic population are known, samples are drawn only once and the sample size is defined in advance. In theoretical sampling, elements are drawn repeatedly and continuously throughout the process (see Chapter 6) until a theoretical saturation has been reached. This sometimes makes it difficult to calculate in advance how many cases you will need (see the expert discussion on this point in Baker and Edwards 2012).

Purposive sampling

Patton (2002) suggests the following variants of purposive sampling:

- Extreme cases are characterized by a particularly long process of development or by the failure or success of an intervention.
- Typical cases are typical of the average or the majority of potential cases. Here, the field is explored rather from the inside, from the center.
- Maximum variation sampling includes a few cases which are as different as possible, for analysis of the variety and diversity in the field.
- Intensity sampling includes cases which have a different intensity of the relevant features, processes or experiences or for which you assume such differences. Either way, you will include and compare cases with the highest intensity or with different intensities.

- Critical cases show the relations under study particularly clearly, or are very relevant for the functioning of a program under study. Here, you often look for advice from experts about which cases to choose.
- Politically important or sensitive cases can be useful for making positive results widely known.

Convenience sampling

Convenience sampling refers to choosing those cases that are most easily accessible under given circumstances. This can reduce the effort in sampling. Sometimes this is the only way to do a study with limited resources of time and with difficulties in applying a more systematic strategy of sampling.

Sampling and Access in Online Research

As in general research, so in online research: the concept of sampling involves a wider population from which you draw a sample, and so your sample of actual participants represents some larger group of potential participants. In online research, however, you may have the problem of a double reality. To take a simple example: if you wish to study trends in book purchasing and you use an online shop like Amazon to find your participants, you will be using a very selective approach – you cannot necessarily equate the clients of this shop with those of (a) offline bookshops or even (b) other shops on the Internet. Thus, in this example, it will be difficult to draw conclusions from a sample (of Internet bookshop users) to a population of book buyers in general. This example entails questions not only of how representative online samples are for real-world populations, but also of access – where you will find participants, how you will contact them, and so on.

Here, we may consider a number of forms of access for an online survey (see below) aiming at a random sample (see Baur and Florian 2009). They are:

- web surveys without scientific claims
- open web surveys with no limitations on who is expected to take part; everyone is invited (through, say, a banner on a website) and may even be able to participate several times over
- self-recruited volunteer panels that address potential participants drawn to a certain issue
- intercept surveys that draw a random sample of all people who visit a certain webpage and then invite the members of that sample to participate in a survey

- list-based surveys that use institutions with a complete e-mail list of its members (e.g. a university with all its students and employees), all of whom are invited to participate (or from which a random sample is drawn)
- mixed mode surveys that use a random sample of the population; members of the sample may then choose between paper and online questionnaires for their participation
- a panel of the population recruited beforehand, drawn from a random sample of the population and where, if members of this sample do not have access to the Internet, the researchers will provide them with access.

These alternatives differ in regard to where you find your population to draw a sample from and also in how far you can realize the idea of random sampling in this context. The population may consist either of the users of a website or Internet portal, or of the general population of a country. In the latter case, you have to take into account the fact that Internet users constitute a selection of the general population: by no means does everyone use the Internet. For example, Internet users tend to be younger than the average population and are more likely to have received a higher education. This may lead to problems of *under-coverage* in your Internet sample compared to the general population: that is, certain groups of the population (e.g. older people or those without children) will be systematically under-represented in your Internet-based sample. At the same time, you may face *over-coverage* as a problem: people who are not in your target group or sample may respond to your questionnaire, sometimes without revealing their identity, which then makes it difficult to exclude them from your dataset. Or some people may fill in your questionnaire several times. Other problems include non-response to certain items/questions, or respondents who drop out, for example, after answering the first group (or page) of questions due to losing interest or suffering from a breakdown in Internet connection.

Another problem of access is what and how much you will really know about your participants if you use people's e-mail address or the nickname they use in discussion groups or chat rooms to identify them. In some cases, you will know no more about them or have to rely on the information they give you about their gender, age, location, and so on. This may raise questions about the reliability of such demographic information and lead to problems of contextualizing the statements in the later interview. As Markham asks: 'What does it mean to interview someone for almost two hours before realizing (s)he is not the gender the researcher thought (s)he was?' (2004, p. 360).

In addition, using an e-mail address as the identifier for participants in your study may raise other issues. Many people use more than one e-mail address or several Internet providers at the same time. In contrast, several people within the same household may use the same computer (see Bryman 2016, p. 191).

Several responses to these problems of sampling have been suggested. One is that not every study needs a sample as closely representative of the 'population at large' as possible: this applies to online research as well as offline (Hewson et al. 2003). Qualitative studies follow a different logic of sampling, and maybe it does make sense to study the shopping trends of amazon.com clients (in our example above) for some purpose. So such a sample may be adequate as long as you refrain from generalizing your results to inadequate populations (e.g. book buyers in general). You need to be careful in generalizing your results to other populations and to reflect on what is adequate and what is not. Finally, for some studies and research questions, it may not be a problem that participants respond repeatedly or use unclear identities as this may represent typical user practices on the Internet. The questions and issues mentioned by Gaiser and Schreiner in this context may be helpful:

> Consider who you want in the study and where you are going to find them. Who is likely to frequent your type of online environment? How can you get participants to participate in your site? How might a researcher best engage a particular population? What technologies do sample participants use or are more likely to use? (2009, p. 15)

In general, as in other forms of survey, you should consider two steps for increasing the response rate. Before you send a questionnaire to potential participants, you should contact them and ask for their permission to include them in your study. You should follow up with non-respondents at least once (Bryman 2016, p. 192).

Summary

Sampling refers to strategies for assuring that you have the 'right' cases in your study. 'Right' means that they allow generalization from the sample to the population because the sample is representative of the population. For example, the results from a questionnaire study with a sample of youth should be able to be generalized to youth in Germany. 'Right' can also mean that you have found and included the most instructive cases in your interviews – that you have the full range of health experiences of homeless adolescents, rather than that your results are valid for youth in Germany in general (see the examples in Chapter 4).

These sampling strategies are a major step in research planning. Some research designs will need one or the other form of sampling. Experiments, control group or double-blind studies need random sampling to be successful. For a qualitative, theory-developing study, strategies of theoretical or purposive sampling are more appropriate. The same is the case for the models of the research process that we discussed earlier: random and similar forms of sampling are more suitable for the process model of quantitative research; theoretical and purposive sampling are better suited to the qualitative research process.

What You Need to Ask Yourself

The questions in Box 7.4 should help you clarify the approach of your study and its design.

Box 7.4   What you need to ask yourself

Research design

1   Do you have a clear idea of which methodological approach you want to apply?
2   Is the form of sampling you want to apply appropriate for reaching the goals of your study and also for reaching the target groups of your study?
3   Do you have a clear idea of which participants to look for and select for your study – and why?
4   Do you know how and where to find them?
5   Is the design you chose adequate for the aims of your study and for the conditions in your field of study?
6   Do the proposal and the timescale cover the main steps of your project?

What You Need to Succeed

To plan your own empirical project, you should take into account the points in Box 7.5 and find answers to the questions that arise. This checklist may help you in planning your own study and also in assessing the existing studies of other researchers.

Box 7.5   What you need to succeed

Research design

1   Write a proposal for your project to indicate which steps it should run through.
2   Develop a timescale for your project in order to ascertain whether you can manage it in the available or given time frame.

(Continued)

3   You should have developed a clear idea about what design means in quantitative and in qualitative research. You should have an idea about the process of your project.

4   You should also have an idea of which resources you need and which ones are available for you to run the study.

What you have learned

- Writing a proposal and developing a timescale are necessary steps in making a project work.
- Research designs differ, in terms of both planning and procedures, between qualitative and quantitative research.
- Quantitative research focuses more on control and standardization of conditions of data collection. Qualitative research is more interested in planning the study in a design.
- Statistical sampling is meant to allow for generalization of the results to the (known) population.
- Sampling in qualitative research is more oriented towards the purposive selection of cases, which in the end will give insights into features of the population.

What's next

Issues of designing social research in a quantitative approach are discussed in the first three books here:

Bryman, A. (2016) *Social Research Methods*, 5th edn. Oxford: Oxford University Press.

Neuman, W.L. (2014) *Social Research Methods: Qualitative and Quantitative Approaches*, 7th edn. Boston: Allyn & Bacon.

Blaikie, N. and Priest, J. (2019) *Designing Social Research*, 3rd edn. Cambridge: Polity Press.

This first book outlines alternatives in qualitative research design, while the second puts this question in a broader framework:

Flick, U. (2018) *Designing Qualitative Research*, 2nd edn. London: Sage.

Flick, U. (2018) *An Introduction to Qualitative Research*, 6th edn. London: Sage. Part 3.

The next source is a comprehensive overview of the field of online research methods:

Fielding, N.G., Lee, R. and Blank, G. (eds) (2017) *The SAGE Handbook of Online Research Methods*, 2nd edn. London: Sage.

The following article is a comprehensive reflection on how to put together a thorough proposal:

Abdulai, R. T. and Owusu-Ansah, A. (2014) 'Essential Ingredients of a Good Research Proposal for Undergraduate and Postgraduate Students in the Social Sciences', *SAGE Open*. https://doi.org/10.1177/2158244014548178.

The following case study can be found in the online resources. It should give you an orientation on how a research design can be developed:

Mendoza, V. (2014) 'Measurement, Tips, and Errors: Making an Instrument Design in Risk Perception', *SAGE Research Methods Cases*. doi: 10.4135/978144627305013519224.

The following article, also in the online resources, clarifies the term 'research design' by comparing three prominent approaches:

Abutabenjeh, S. and Jaradat, R. (2018) 'Clarification of Research Design, Research Methods, and Research Methodology: A Guide for Public Administration Researchers and Practitioners', *Teaching Public Administration*, 36(3): 237–58.

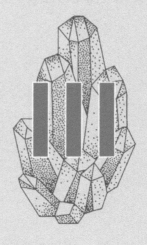

METHOD SELECTION

Part I of this book was designed to help orient you towards doing your research project. Part II guided you through key steps in the early phases of planning the project itself, whether it be qualitative or quantitative. Part III focuses on selecting methods for doing your study.

Your project plan will become more concrete as you decide which methods you wish to apply. Chapter 8 outlines the decisions you need to make in the research process when choosing your method, the form of sampling or the type of research. This should lead you to the reflective stage, halfway through the process, in which you consider again the implications of your research plan, before you start to actually apply concrete methods and begin to work with data.

Every method has its limitations and so it can be fruitful to combine methods. Chapter 9 outlines ways to do this through triangulation and mixed methods research.

Research Methodology Navigator

Orientation

- Why social research?
- Worldviews in social research
- Ethical issues in social research
- From research idea to research question

Planning and design

- Reading and reviewing the literature
- Steps in the research process
- Designing social research

You are here in your project

Method selection

- Deciding on your methods
- Triangulation and mixed methods

Working with data

- Using existing data
- Collecting new data
- Analyzing data

Reflection and writing

- What is good research? Evaluating your research project
- Writing up research and using results

8

DECIDING ON
YOUR METHODS

How this chapter will help you

You will:

- understand the series of decisions you will be required to make during the research process
- appreciate that selecting a specific method of data collection is an important decision – though only one of many – and
- see how your decisions concerning methods are related to more general issues concerning (a) your research, (b) the conditions in the field, and (c) available knowledge about the issue.

Decisions in the Research Process

In Chapter 6, I outlined the steps involved in quantitative and qualitative research processes. In the chapters that follow, the most important methods will be discussed in more detail (see Chapters 11 and 12). Both qualitative and quantitative research entails a series of decisions you will need to take – from defining your research question, to collecting and analyzing data, and finally to presenting your results. Each decision will have implications for subsequent stages in your research project. Figure 8.1 gives a first orientation of process that is then outlined in more detail.

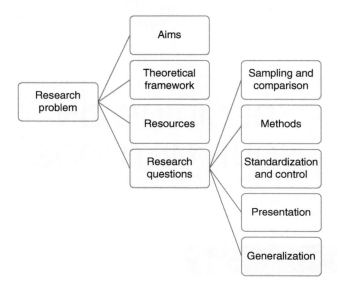

Figure 8.1   Decision process in a research project

Overviews of methods in social research rarely provide much advice on the choice of specific research methods. This chapter aims to make good that lack.

Decisions in Planning a Quantitative Study

The first decision you need to make concerns the selection of a research problem. This will have major implications for the subsequent procedures.

Selecting the research problem

Bortz and Döring (2006) have formulated a number of criteria for evaluating research problems or ideas for studies. These can be used to inform your decision. Their criteria are:

- Precision in the formulation of the problem: how vaguely or exactly is the idea articulated? How clear are the concepts on which it is based?
- Can the problem be studied empirically? Can the ideas be addressed empirically, or are they based on religious, metaphysical or philosophical content (e.g. concerning the meaning of life)? And how likely is it that sufficient potential participants can be reached without excessive effort?
- Scientific scope: has the topic already been studied so comprehensively that no new insights can be expected from further investigations?
- Ethical criteria: would the study violate any ethical principles (as discussed in Chapter 3) or, seen the other way round, is the study ethically justified?

These criteria will help both to assess your research ideas and to justify your selection. Let us take as an example a study of the perception and effects of a change in the school system – cutting down the number of high school years and compensating for this by extending the number of classes per day. This is a rather clear issue with clear concepts (number of classes, time at school in years, effects on those concerned at school). As this is a new phenomenon (it was introduced in Germany around 2010), there is not much research about this. There are no reasons why this study should be ethically problematic.

Deciding on the research problem in a quantitative study

Your decision at this point concerns the research problem as such and those aspects of it that will be in the foreground of your study. They should be oriented to your interests and to how far you can formulate them empirically. Furthermore, you should assess whether or not the existing knowledge about the problem is sufficient for doing a quantitative study and whether you will be able to access a sufficient number of participants. Decisions at this stage will subsequently have an influence on your methodological decisions.

Aims of the study

Quantitative studies usually aim at testing an assumption that has been formulated in advance in the form of a hypothesis. Here, the aim will be to assess the connections between variables or to identify the causes of specific events. One should be careful here not to assert a relationship or causal link between variables without evidence to justify the assertion. Thus, there will need to be a strong emphasis, when planning and designing the study, on standardizing as many conditions as possible and on defining variables.

It is important to distinguish between independent and dependent variables (see also Chapter 7). The 'causing' condition is labeled the 'independent variable' and the consequences the 'dependent variable'. For example, an infection will be the cause of certain symptoms. The symptoms occur due to the infection. Thus, they depend on the existence of the infection and are therefore treated as dependent variables. The infection is independent of the symptoms: it just occurs. It is therefore treated as the independent variable. Sometimes this relation is not so immediate. Other factors may play a role – for example, not everyone exposed to an infection falls ill; some people have dramatic symptoms due to an infection; and others have less dramatic symptoms despite the same infection. Thus, one must suppose that other variables are relevant. These are called intervening variables. This term is a label for those other influences on the connection between independent and dependent variables. In our example, the social situation of the infected persons could be such an intervening variable. (For example, socially disadvantaged people may be more likely to develop strong symptoms than those in better living conditions.) Testing the relationship between the independent and dependent variables can be your aim – in our example, the relation between infection and disease (as seen from the symptoms) and thus the identification of the infection as the cause of the disease. In this example, it will be important to control the influence of the intervening variable.

Another aim of a quantitative study may be to describe a state or situation – for instance, the frequency of a disease in the population or in various subpopulations. Such studies are known as 'population description studies', as distinct from 'hypothesis testing studies' (Bortz and Döring 2006, p. 51). When the state of research and the theoretical literature are not sufficiently developed for you to formulate hypotheses that you can test empirically, you may first conduct an exploratory study. In this kind of study, you may develop concepts, explore a field and end up by forming hypotheses based on your exploration of the field.

Coming back to our example, the aim could be to identify the effects of the change in length of schooling. Reducing the number of years of school attendance is the independent variable (fewer years with more classes) and the effects on the individual student are the dependent variables (e.g. more stress). Intervening variables may be social class, ethnic background and the like.

Deciding on the aims of a quantitative study

Your decision on the type of study you will pursue should be determined by your research interest and the state of the research before your study. The relevant question at this stage is: how far is your decision determined by your issue and field of study? Or is it (mainly) influenced by your general methodological orientation?

Theoretical framework

Theoretical questions can become an issue for decisions on a number of levels. You should take as your starting point a specific theoretical model of the issue that you intend to study. Often, a number of alternative models will be available. For example, if your study concerns coping behavior in the case of a disease, there are several different models of coping behavior to select from. Moreover, the study may be planned within the framework of a general theoretical model or research program. For example, in the case of studying coping behavior, one might adopt a rational choice approach. In our example from education, the availability of schools with fewer years and those with more years may allow students to make a rational choice between one and the other type of schooling.

Deciding on the theoretical framework of a quantitative study

If you decide to use a theoretical framework, this will have a number of methodological consequences. Deciding on a specific theoretical model of the research issue will set the frame for operationalizing the relevant features of the issue in your study. A relevant question in this context is how far the theoretical framework is compatible with your research question, or with the issue. These decisions should be oriented to the issue under study and the field in which you study it.

If, for example, you choose to study quality of life according to one of the theoretical models developed about people living on their own in relative health and independence, this will lead to such operationalizations as a question about people's ability to walk certain distances, for example. If you want to study this issue (quality of life) in a nursing home with frail old people, you will have to consider whether such questions and the theoretical models in the background are appropriate for this context.

Decisions in designing a quantitative study

Bryman (2007) discussed the role of the research question after interviewing various researchers. He found that either a particularistic view was expressed – the research question should guide all other elements of a study – or a universalistic discourse – the research question is a part of the process and it is often the research program which determines what is studied within a project.

Formulating the research question

For the success of any study, it is important to limit the chosen research problem to a research question that is manageable. For example, if you are interested in the research issue 'The health of senior citizens', this is not yet a research question, as it is too vast and vague. To turn it into a research question, you will have to focus on parts of the problem formulation. Which aspect of health do you want to study? What kind of senior citizens are in the focus of your study? What is the link between health and older people? Then you may arrive at a question like: 'Which factors delimit the autonomy of people over 65 with depression living in big cities and in rural contexts?' Here, the elements of your research question are clearly defined and you can start to consider how you will do your sampling and data collection in order to approach this question empirically. If you want to do a quantitative study, you should reflect on whether there will be enough people for you to address and whether they will be capable of filling in a survey, etc., and how this survey will cover issues of autonomy, of limitations, of living conditions (city, countryside) and of depression (see Chapter 4 for more examples of and distinctions between good and bad research questions). If we come back to our example from education, we can, for instance, formulate the research question: How does the condensation of schooling (eight instead of nine years and more classes) contribute to social inequality among students?

Deciding on the research question of a quantitative study

Your decision regarding the research question will have implications for (a) what will become the issue you study, (b) which aspects you will omit, and (c) which methods will apply to your study. At this stage, it is important that the formulation of your research question helps you to orient your research. It is also important how far your research question is helpful in stimulating new insights about your research issue, so that your study does not simply reproduce knowledge already available in other research.

Resources

A key factor is the cost of a study. Without detailed knowledge about the project, an estimate of the cost will be difficult to make. In general, the higher the methodological standards, the greater the cost. Denscombe (2007, p. 27) mentions in this context that commercial survey institutes in Britain inform their customers that for a certain price a specific level of exactness in measurement and sampling can be expected, and that higher levels will incur higher prices. Accordingly, Hoinville et al. state in the case of sampling that 'In practice, the complexity of the competing factors of resources and accuracy means that the decision on sample size tends to be based on experience and good judgment rather than relying on a strict mathematical formula' (1985, p. 73). In our example from school research, we need as resources a questionnaire and access to a sufficient number of participants to create groups for comparisons of their evaluations of the change.

Decisions concerning the use of resources in a quantitative study

Your decision in this context generally refers to weighing your available resources (money, time, experience, wo/manpower) against methodological claims (of exactness and scope of the sampling, for example), so that you can make your project work with realistic claims.

Sampling and building of comparative groups

Sampling in quantitative research mostly rests on concern for the representativeness of the studied persons, situations, institutions or phenomena for the wider population. Often, one will construct comparative groups so that they match each other as far as possible (e.g. the study group will be constructed like the control group as much as possible). The aim here is to control and standardize as many features of the group as possible; then differences between groups can be traced back to the variable that you are studying. The most consistent approach is random sampling, in which allocation to the study group or the control group is done randomly (see Chapter 7). However, strictly random sampling is not always the best or most appropriate way. Depending on your issue and field of study, quota or cluster sampling may be more appropriate (see Chapter 7) as the focus of strictly random sampling may be insufficiently specific. In our example from school research, we may try to do some kind of cluster sampling. This means first including various schools from different parts of a big city, or from urban and rural areas; and then including students randomly in each school or again in clusters (girls, boys, with or without migrant backgrounds, with different social backgrounds or the like).

Deciding on sampling and comparison in a quantitative study

Your decisions here relate to the question of the appropriateness of a specific form of sampling – does it take the specific target groups of your study sufficiently into account?

Decisions in doing a quantitative study

Decisions concerning methods need to be made on a series of levels.

Methods

The first decision concerns the character of the data you wish to work with. Ask yourself whether you can use existing data (e.g. routine data on health insurance) for your own analysis. Here, one has to consider the question of accessibility of data (e.g. not every health insurance company is ready to make its data available for research purposes). Sometimes data protection issues form obstacles. There will also be questions about the suitability of the data – in particular, you should check whether the issue you are interested in is indeed covered by the data, and whether the way the data are classified permits the necessary analysis (see Chapters 10, 11 and 12).

Next, you should decide between a survey and an observation. For example, when collecting data about the relevant phenomena, are you interested more in knowledge and attitudes or in practices?

The next decision to be made is whether for data collection you use an existing instrument or develop a new one. Advantages of the first option are that these methods are mostly well tested and that you can more easily link your data to other studies. For example, in quality of life research, existing questionnaires are often used; similarly, in attitude research, interactions are often analyzed with available inventories in observations. However, you should check whether the existing instrument covers the aspects that are relevant for your own study and whether it is appropriate for your specific target group.

Developing your own instrument enables you to adapt it to the concrete circumstances of your study. In this case, you should reflect on whether the existing theoretical or empirical knowledge is developed enough for you to formulate the 'right' questions or observational categories. Finally, pre-testing and checking the reliability and validity (as discussed in Chapter 13) of the instrument are necessary before you can actually apply it.

In analyzing quantitative data, existing statistical packages like SPSS are most often used. Here, you should decide which kinds of relational analyses are best for answering your research question. Also, you should check in advance which tests you need to apply to your data – for example, plausibility checks (are there contradictory responses in the dataset, like 20-year-old pensioners?) or checks for missing data (see Chapters 11 and 12).

In our example, we would develop a questionnaire and use a package like SPSS for analyzing correlations between answers concerning stress experiences related to this condensed kind of schooling and social background features (social status, for example).

Deciding on methods in a quantitative study

Your decision here is between using existing data or instruments and collecting your own data, perhaps with instruments developed specifically for your study. This decision should be related to your research question, the conditions in the field and existing knowledge about the issue under study. Finally, your decisions may be a matter of your resources – such as the time available or the researchers' methodological skills.

Degree of standardization and control

Quantitative research is based on (a) standardizing the research situation and the research procedures and (b) controlling as many conditions as possible. In most cases, variables are defined that are linked in hypotheses for testing these connections. Analytic units are defined (for example, every patient attending a general practitioner's surgery). Concrete measures are defined for the single variables (for example, the time that each patient waits before he/she is called into the treatment room to see the doctor). These are defined before entering the field and are then applied to every case in an identical way. This is intended to ensure the standardization of the research and the control over the conditions in the research situation as far as possible.

Deciding on standardization and control in a quantitative study

Your decisions in this context refer to how far you can or should advance with standardizing and controlling. Experimental studies are most systematic when seen from a methodological point of view. However, they cannot be applied to every field and every issue. Other study forms are less standardized and controlled, but can more easily fit the conditions in the field under study. The decisions you take in this context concerning more or less standardization and control should be defined both by the conditions in the field and by the aims of your study.

Decisions in *transferring* a quantitative study

Once you have done the core of your study – planned and designed it, collected and analyzed the data – you will reflect on how to transfer the study and its findings on two levels: how far you can generalize your findings and how to present what you find.

Generalization

Generalization usually involves inference from a small number (of people in the study) to a larger number (of people that could have been studied). Accordingly, generalization may be seen as a numerical or statistical problem. It is closely linked to the question of the (statistical) representativeness of the sample that you have studied for the population that you assumed (as discussed in Chapters 7 and 13). Note here that 'population' does not necessarily refer to the whole population of a country; in many studies, the term will refer to rather more limited basic populations. In our example from school research, the question is how far the results can be generalized to students in Berlin or in Germany or only to more limited parts of these populations.

Deciding on generalization in a quantitative study

Deciding on the specific target population for the purposes of generalization will have consequences for the research design and for the methods that you apply. This decision should be driven by the aims of your study in general and by conditions in the study field. The general question here is how appropriate the intended generalization is to (a) the study topic, (b) the field, and (c) the participants.

Presentation

Who do you want to address with your research and its results? What will be the audience and target group when it comes to presenting your findings? Here, we can distinguish between (a) academic, (b) general, and (c) political audiences. If your study will in the end be presented in a thesis (a master's thesis, for example), it will be more important that you demonstrate specific methodological competences than that the results draw the attention of the general public to a social problem. If your research and its results are intended to have an influence on political decision-making, their presentation will need to be concise, easily understandable and focused on the essential results (see Chapter 14). In our example, the results of our study into correlations between the reduction in school years and social inequality should be elaborated on for publication in a scientific journal in education. They should also be condensed in a way that permits presenting them to political or administrative decision-makers in order to produce some impact on the planning of school programs.

Deciding on how to present a quantitative study

Your decisions at this stage concern the kinds of information you should select for the audience you want to address. A second issue is what style of presentation is appropriate for this purpose.

Decisions in Planning a Qualitative Study

This process of decision-making in a quantitative study can be applied in an adapted way to illustrate the development of a qualitative study, too.

Selection of the research problem

Many factors affect the choice of research problem in qualitative research. It may be that the theoretical literature or empirical research to date is lacking in some way. Alternatively, one might choose a qualitative approach because the participants in question would be difficult to reach through quantitative methods. Another factor influencing the choice may be that the number of potential participants (e.g. people with a specific but rare diagnosis) is small (though not too small), or one may wish to explore a field to discover something new. The decision over the choice of problem will also involve a consideration of ethical issues (discussed in Chapter 3). If we come back to the example from school research used before, we can address the phenomenon from a different angle now. Whereas the question-naire study reveals the distribution of attitudes towards this issue of school reform by using a sample according to formal (statistical) criteria, a qualitative study could address the individual experiences of students selected for a specific combination of features.

Deciding on the research problem in qualitative research

Questions you should focus on here are: What is new about the problem under consideration? Which aspects of it can be researched empirically and discovered? What are the limitations of existing research? And can a sufficient number of participants be accessed? Your decisions at this stage will influence the methodological steps that you take later in the project.

Aims of the study

According to worldviews in social research (see Chapter 2), qualitative research will be more interested in understanding subjective experiences and social processes in the making than in collecting 'facts'. Maxwell (2005, p. 16) has distinguished between different types of research aims. There are (a) personal aims such as completing a master's or doctoral thesis; (b) practical aims such as finding out whether a specific program or service works; and (c) research aims concerning the desire for general knowledge on a specific issue.

Qualitative studies often have the aim of developing grounded theory according to the approach of Glaser and Strauss (1967). However, this is an ambitious and demanding aim. If you are writing a bachelor degree thesis, this aim may well be unrealistic: you may not have the time or experience required. It may be more realistic to aim instead to provide a detailed description or evaluation of some ongoing practices. Overall, qualitative research may aim to provide description or evaluation or to develop theory. In our example, we should try to unfold the variation in experiences of the schooling conditions by interviewing a variety of students with differing backgrounds.

Deciding on the aims in a qualitative study

Your decision here concerns the aims you can realistically pursue within your study.

Theoretical framework

In qualitative research, it may be that you do not use a theoretical model of the issue under study to provide a starting point for determining the actual questions you use (or those you ask in an interview). Nevertheless, studies should be related to previous theoretical and empirical work on the issue in question. The current state of extant research should influence your subsequent methodological and empirical procedures. In qualitative research, there may be a number of frameworks for studying an issue – for example, it may be that you can analyze either (a) subjective views and experiences or (b) interactions related to the topic in question. In our example, we decide to focus on subjective views on the part of students concerned with structural changes in the school system.

Deciding on the theoretical framework in a qualitative study

When you decide on the research perspective and the substantive points of the research, you will in effect be committing yourself to proceeding in certain ways. In doing so, you should take as your points of reference the knowledge available to you and the conditions in your field of study.

Decisions in designing a qualitative study

In qualitative research as well, research should be planned while developing a research design. This will again include several aspects.

Formulating the research question

We can distinguish between (a) research questions where the answers focus on the possible confirmation of an assumption or a hypothesis, and (b) questions designed to discover new aspects. Strauss (1987) calls the latter 'generative questions'. He defines them as 'questions that stimulate the line of investigation in profitable directions; they lead to hypotheses, useful comparisons, the collection of certain classes of data, even to general lines of attack on potentially important problems' (1987, p. 22).

In qualitative research, Maxwell (2005) has proposed alternative distinctions. He distinguishes first between generalizing and particularizing questions, and second between questions that focus on distinctions and those that focus on the description of processes. Generalizing questions place the issue under study in a wider context – for example, as the biography of a person or group could be understood against the background of a political crisis. Particularizing questions foreground some specific aspect – for example, a specific event, such as the onset of illness. Questions focusing on distinction address differences in the knowledge of people – say, several patients' differences in knowing about their illness. Questions focusing on describing a process look at how such knowledge develops in a group of patients in the progress of their illness. In our example from school research, we would focus on how subjective experiences of a shorter school life have turned into varied forms of knowledge about this process.

Deciding on the research question in a qualitative study

Selecting a research question entails decisions both about what exactly you will be studying and what you will be excluding from your study. This will have implications subsequently for your choice of methods for data collection and analysis.

Resources

When one is elaborating on a research design, the resources required (time, personnel, technologies, competencies, experiences) are often underestimated. In research proposals, it is common to see a mismatch between work packages envisaged and the personal resources that have been requested. To plan a project realistically, you need to assess accurately the work to be undertaken.

For example, for an interview lasting 90 minutes, it is recommended that you allow an equivalent amount of time for recruiting the interviewee, organizing the appointment, etc. To calculate the time needed for transcribing the interviews, estimates vary according to the degree of exactness of the transcription rules that are applied. Morse (1998, pp. 81–2), for example, suggests that for fast-writing transcribers, the length of the tape containing the interview recording be multiplied by a factor of four. If checking the finished transcript against the tape is also included, the length of the tape should be multiplied by six. For the complete calculation of the project, she advises doubling the time, to allow for unforeseen difficulties and 'catastrophes'. Example plans for calculating time schedules or empirical projects may be found in Marshall and Rossman (2006, pp. 177–80) and Flick (2018b). In our example, we should plan interviews with the students and their transcription and allow enough time for this. To achieve good data, it would also be helpful to plan time before the actual interviews for interview training to improve the interviewers' skills as a resource.

Decisions concerning the use of resources in a qualitative study

When it comes to decisions in this context, you should, above all, take the relationship between the available resources and the aims or the planned efforts of the study into account. This should help you to ensure that the data you collect is not too complex and differentiated for you to be able to analyze them in the time available.

Sampling and building of comparative groups

Decisions about sampling in qualitative research above all refer to persons or situations in data collection. Sampling decisions become relevant again for the parts of the collected material that you will address with extended interpretations when analyzing your data. In the presentation of your research, sampling concerns what you present as exemplary results or interpretations (see Flick 2018a, p. 185). Sampling decisions here are not normally taken according to abstract criteria (as in random sampling), but rather according to substantive criteria referring to concrete cases or case groups.

A major task for the sampling decision is to build comparative groups. Here, you need to decide at what level you want to do your comparisons. For example, will your focus be the differences and similarities between people and institutions, or between situations and phenomena? In our school example, we should try to include students from groups as varied as possible – for example, different social backgrounds, migration experiences, high or lower performers in school, etc.

Deciding on sampling and comparison in a qualitative study

Here, you make decisions about the persons, groups or situations that you will include in your study. The decisions should be oriented to the relevance of who or what you select for your study. They should also be oriented to having sufficient diversity in the phenomena you study. If that is your topic, you should look for people with a *specific* illness experience (and not just people who are sick in some way or another). At the same time, your selection should provide for some diversity, for example people living in different social circumstances with this illness experience and not just people living in the same conditions.

Decisions in doing a qualitative study

Here again, we make decisions between available alternatives, which involve several issues.

Methods

The central distinction here is between (a) direct analysis of what occurs and (b) analysis of reports about what has occurred. The former will involve (participant) observation or interaction studies. In the latter case, you will work with interviews with or narratives of the participants. You can decide between different degrees of openness or structure: data collection can rely on questions formulated in advance or on narratives; observation is either structured or open and participant. The analysis of data can be oriented on categories (sometimes defined in advance) or on the development of the text (of the narrative or of the interaction protocol: see Chapters 11 and 12). In our example, we would focus on interviewing students about being able to address longer processes of change in the past as well and would apply coding to the data which focuses on developing the codes from the material (see Chapter 12).

> ## Box 8.1 Research in the real world
>
> **A student's research decision regarding methods for studying a professional concept in hospice work**
>
> In her master's thesis, Franziska Wächter (2019; see Box 7.3 in Chapter 7) reflected on and discussed a number of methodological alternatives for doing interviews. Before deciding on the episodic interview, she discussed the expert interview (as her interviewees have developed an expertise in their field), the focused interview, the problem-centered interview and the narrative interview (as she was interested in narratives on professional practice). She then decided to use the episodic interview as it combines a number of features of the other methods – narratives, the focus on practices and problems – and allowed her to approach the interviewees as experts too. So her reflections led to a weighing up of the advantages and disadvantages of several methods and, ultimately, a reflected decision made on a specific method.

Deciding on the methods in a qualitative study

Your decision here concerns the level of data (report or observation) and the degree of openness or structure in the data collection and analysis. Other points of reference – besides your research question and the particular conditions in the field – will be the aims of your study and the resources available.

Degree of standardization and control

Miles and Huberman (1994, pp. 16–18) distinguish between tight and loose research designs in qualitative research. Tight research designs involve narrowly restricted questions and strictly determined selection procedures. The degree of openness in the field of investigation and the empirical material will remain limited. The authors see such designs as appropriate when researchers lack the experience of qualitative research, when the research operates on the basis of narrowly defined constructs or when it is restricted to the investigation of particular relationships in familiar contexts. In such cases, they see loose designs as a detour to the desired result. Loose designs are characterized by less defined concepts and have, in the beginning, hardly any fixed methodological procedures. Tight designs make it easier to decide which data are relevant for the investigation. They make it easier to compare and summarize data from different interviews or observations. In our example, we would develop a loose design for adapting more appropriately to the individual cases of the interviewed students.

Deciding about standardization and control in a qualitative study

In qualitative research, standardization and control play a minor role compared to the case in quantitative research. However, you can try to reduce the variety in your material and try to focus your approach as far as possible (a tight design). Alternatively, you can reduce standardization and control if you decide on a more open and less defined approach (a loose design). Both have their advantages and disadvantages.

Decisions in *transferring* a qualitative study

Once you've done the core of your study – planned and designed it, collected and analyzed the data – you must reflect on how to transfer the study and its findings on two levels: how far you can generalize your findings and how you will present what you find.

Generalization

With qualitative research, aims may vary – for example, between (a) providing a detailed analysis of a single case in as many of its aspects as possible, (b) comparing several cases, and (c) developing a typology of different cases. The generalization involved is likely to be theoretical rather than numerical. The important consideration is more likely to be the diversity of cases considered or the theoretical scope of the case studies than the number of cases included. To develop a theory can be a form of generalization on various levels as well. This theory can refer to the substantive area that was studied (e.g. a theory of trust in counseling relationships). Generalization can be advanced by developing a formal theory focusing on broader contexts (e.g. a theory of interpersonal trust related to various contexts). This distinction between substantive theory and formal theory was suggested by Glaser and Strauss (1967). In our example, we could develop a grounded theory of experiences with shortening school processes from the view of the students that are subject to this process. We could then look for other empirical approaches and materials to see whether we can develop a more formal theory of time compression and acceleration in the current life course.

Deciding on generalization in a qualitative study

Your decision concerning the kind of generalization that you aim for will have implications for planning your study and, in particular, for your selection of cases. This decision should take account of the aims of your study and, at the same time, the question of what is possible in your field of study. It should also take account of the circumstances of the possible participants. More generally, the question arises of how appropriate the type of generalization that you are aiming at will be for the field under study.

Presentation

Finally, you should take into account in your planning the question of presentation. Will the empirical material form the basis for an essay or a narrative with a more illustrative function? Or is your aim to provide a systematic study of the cases under study and the variations between them? You need to consider here the assessment criteria that will be applied to your thesis. In general, the question will be how to relate concrete statements and evidence to more general or deepening interpretations so that your inferences are substantiated in a clear and convincing way. (Questions of presentation will be discussed in more detail in Chapter 14.) In our example, it will be necessary to sufficiently document the variety of views in the interviews when writing an article, for example.

Deciding on how to present a qualitative study

You should carefully decide how you will present your research. It is important that you not only present some results, but also make transparent how you entered the field, how you got in touch with the relevant people, and how you arrived at gathering the data you

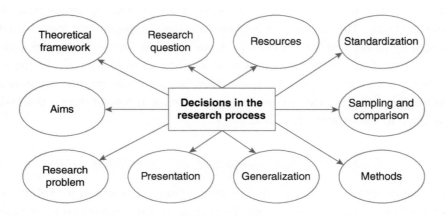

Figure 8.2 Decisions in the research process

needed for your analysis. It is also important that your readers have access to the way you gathered your data and to how you analyzed them. The path from the original data to the more general (comparative, analytic) statements and conclusions should be elucidated with examples in the methods section of your report and with enough original material (quotes) in the results section. Illustrations with sample materials and with charts or tables can be very helpful here.

Decisions within Quantitative and Qualitative Research

Planning a research project involves making a series of decisions that serve to foreground some aspects and exclude others. The decisions require consideration of interrelated questions concerning your field of study, the issue to be researched, the theoretical context and the methodology involved.

The decisions discussed in this chapter so far are summarized in Table 8.1. These decisions (displayed in Figure 8.2) will inform the shaping of the research design and of the research process in its further steps.

Table 8.1 Decisions within quantitative and qualitative research

Decision	Quantitative study	Qualitative study
Research problem	Enough existing knowledge? Enough participants accessible?	What is new about the problem? Enough participants accessible?
Aims	Study type Research interest State of the research	Knowledge interest and practical aims?
Theoretical framework	Model as a basis for operationalization?	Research perspective taken?
Research questions	Helpful for working with the research problem?	Delimiting the research question?
Resources	Methodological claims in relation to money, time, etc.	Is it possible to analyze the collected data within the given time frame?
Sampling and comparison	Appropriateness of sampling: are your target groups taken sufficiently into account?	Diversity of the phenomena in the sample?

(Continued)

Table 8.1   (Continued)

Decision	Quantitative study	Qualitative study
Methods	Your own or existing methods? New or existing data?	Openness and structure of the data coming from the methods?
Standardization and control	Limits of standardization in the field	Comparability of the differences in groups or fields?
Generalization	How far is a generalization intended?	How far can a generalization be intended?
Presentation	Condensation of the essential aspects?	Understandability and transparency of the procedures?

Deciding between Qualitative and Quantitative Research

The following factors provide starting points for deciding between qualitative and quantitative approaches to your empirical project:

- The issue you study and its features should be your major points of reference for such a decision.
- Theoretical approaches have implications for selecting your methodological approaches.
- Your concrete research question will play a major role in defining how you focus your issue conceptually and how you cover it empirically.
- Methodological decisions between qualitative and quantitative methods and designs should be derived from the points of reference mentioned above. They should not be based simply on the belief that only one or the other version of social research is scientific, acceptable or credible.
- A main reference should be the resources available. (With regard to time, however, note that doing a qualitative analysis consistently and carefully in most cases takes as long as a quantitative study.) Your own methodological knowledge and competences are included here under 'resources'.

Overall, your decision between qualitative and quantitative methodologies should be driven more by your research interest and the features of your issue and field of study than by prior methodological preferences. If it is not possible to decide unequivocally between the two approaches, there may be a case for using a combination of the two (as discussed further in Chapter 9).

Deciding between Doing Research On site or Online

Edwards et al. (2013, p. 250) discuss the following points: (1) digital social research as a surrogate for traditional ways of doing research – it can replace them as it can provide insights which other sources of data cannot; (2) digital research can be an augmentation of traditional research methods, for example by helping to reach populations for interview studies that would not be accessible otherwise; (3) digital research can be a re-orientation of social research – as social media communications, for example, raise new questions that ask for different methods and thinking about research.

Furthermore, we can distinguish between using the Internet and social media as a tool – for example, for finding participants for an otherwise traditional interview study – or doing research online – for example, by doing online interviews. Most quantitative and qualitative approaches are now transferred to online forms – from online surveys to virtual ethnographies (see Chapter 11 for this). You may, for example, conduct a survey, not by mailing out copies of a questionnaire and waiting for them to be returned by post, but via the Internet. You may also consider doing your interviews online instead of face to face.

Deciding on Specific Approaches to Research

Textbooks on social research often provide little help concerning the choice between specific methods for a project. Most books treat each particular method or research design separately and describe their features and problems in isolation. In most cases, they fail to provide a comparative presentation, either of methodological alternatives or of the bases for selecting methods appropriate for the research issue in question.

In medicine or psychotherapy, it is usual to check the appropriateness of a certain treatment for specific problems and groups of people. This raises the question of 'indication': one asks whether a specific treatment is 'indicated' (i.e. is appropriate) for a specific problem in a specific case. Similarly, in social research we can ask when (in terms of, for example, the research question, field and issue) qualitative methods are indicated and when quantitative methods are indicated instead. For example, it is common practice to study 'quality of life' for people living with a chronic illness. You will find a number of established instruments to measure quality of life (e.g. the SF-36) which are regularly applied for measuring the quality of life of different populations. The question of indication becomes relevant if you want to use this instrument to study the quality of life of, say, a population of very old people living in a nursing home who are suffering from a number of diseases (and not just one) and are somewhat disoriented. Then you will see that this well-established instrument finds its limits in the specific features of this target group: as Mallinson (2002) has shown, the application of this instrument is not at all clear for such a population and the concrete situation in which they are living. The research question – 'What is the quality of life of old people with multiple morbidity in nursing homes?' – is an appropriate one. However, to answer it you may need other methods than the established SF-36 due to the concrete conditions under which you will study it. For this population, a different method (e.g. an open interview) may be better indicated rather than the common method. This may be different if you study the question of quality of life for a more general population. Then there may be no need to use very open methods and to start from a very open approach in order to develop theories and instruments. Here, enough knowledge about the issue and the population is available to apply standardized and well-established methods. Table 8.2 illustrates this comparison diagrammatically (for more details about this in qualitative research, see Flick 2018a, Chapter 29).

Table 8.2   Indication of research methods

Psychotherapy and medicine			Social research		
Which disease, symptoms, diagnosis, population…	**indicate**	…which treatment or therapy?	Which issue, population, research question, knowledge of issue and population…	**indicate**	…which method or methods?
1 When is a particular method appropriate and indicated?					
2 When is a particular combination of methods appropriate and indicated?					
3 How do you make a rational decision for or against certain methods?					

Reflection Halfway through the Process

Before considering in detail the most common methods available for doing a research project, you might be advised to step back for a moment and reflect on the process of planning so far. Table 8.3 provides a list of guideline questions for examining the consistency and adequacy of your planning to date. Reflecting on these questions will provide a solid foundation for choosing your methods for collecting and analyzing your data.

What You Need to Ask Yourself

The questions in Table 8.3 address the context of choosing and using methodological approaches in your study. The questions in Box 8.2 turn the focus on which methods to select and why.

Table 8.3 Guideline questions for reflecting on your own research project

Issue	Guideline questions	Relevant aspects
Relevance	What is your study important for?	• What theoretical and empirical progress of knowledge do you expect from your results? • What practical relevance do you see for your results?
Clarity	How clearly is your study conceptualized?	• How clear are the aims of your study? • How clearly is the research question formulated?
Background knowledge	What are the bases of your study and of doing it?	• Did you check the current state of research and knowledge on the research issue? Do both justify a quantitative, hypothesis-testing study? Or is there enough of a gap to justify a qualitative study? • Which methodological skill do you have for carrying out the study?
Feasibility	Can the study be realized?	• Are your resources (e.g. time) sufficient for conducting the study? • Did you clarify access to the field and to participants? How likely is the possibility of organizing this access? • Are there (enough) people who can answer your questions?
Scope	Is the approach planned too narrowly?	• Will you include cases, groups, events, etc. in sufficient diversity? • Will the response rate and the readiness to participate be great enough? • Which generalization can you achieve with your results?
Quality	Which quality claims can be formulated for the results?	• Will you be able to apply the methods consistently? • Will the participants' statements be reliable? • Will the data be sound enough to do the intended analysis with them?
Neutrality	How can you avoid biases and being one-sided?	• Can you approach the field and the participants in an unbiased way, even if you do not share their points of view? • Can you avoid acting for or against certain participants in a one-sided way? • Can you accept the limits of your methodological procedures?
Ethics	Is your research ethically sound?	• Can you proceed in your research without deceiving the participants or doing harm to them? • How can you guarantee anonymity, data protection and confidentiality?

Box 8.2   What you need to ask yourself

Choosing your method

1   Can you spell out the reasons why you selected this specific methodological approach?
2   Is this approach suitable for answering your research question?
3   Are the single decisions you take consistent if you look at them in the context of the other decisions, for example does the method of data collection fit your way of analysis and vice versa?
4   Do you see the plan of your research in the line of your decisions?
5   Can you explain this plan and your decisions to other people?

What You Need to Succeed

To select your concrete methods, take the points in Box 8.3 as an orientation. These questions may be relevant in planning your own project and in assessing the existing studies of other researchers.

Box 8.3   What you need to succeed

Choosing your method

1   How far does your methodological approach fit the aims and theoretical starting points of your study?
2   How far does your methodological approach fit the worldviews behind your research?
3   Do your methods of data collection fit those of analyzing your data?
4   Do your data fit the level of scaling and the calculations you will do with them?
5   Are the methods of analysis to be used appropriate to the level of complexity of the data?
6   What implications do the methods (from sampling to collection and analysis) selected have for the research issue and for what is covered in it?
7   Is your decision about certain procedures grounded in your issue and field of study or based on your methodological preferences?
8   Have you assessed which approach(es) is (are) 'indicated'?

What you have learned

- In doing social research, whether qualitative or quantitative, you face decisions at each of the steps outlined in Chapter 6.

- The decisions are interrelated. The method of data collection should be taken into account when you decide on how to do your analysis.
- The aims of your research – for example, who you want to reach and maybe convince with your results – and the framework conditions (e.g. the resources available and the characteristics of the people or groups and fields you are studying) also play a role.
- Decisions between qualitative and quantitative approaches should be driven by your study topic and by your resources.
- The same applies to the decision of whether to conduct your project online or not.
- Checking the 'indication' of research methods provides a starting point for the decisions discussed in this chapter.

What's next

The first text here takes an integrative approach to the selection of research methods:

Bryman, A. (2016) *Social Research Methods*, 5th edn. Oxford: Oxford University Press.

The following book is about how to plan qualitative research, with a strong focus on comparing several methodological alternatives and selecting the appropriate ones:

Flick, U. (2018) *An Introduction to Qualitative Research*, 6th edn. London: Sage.

This next text addresses the selection of methods from a design perspective:

Panke, D. (2018) *Research Design and Method Selection: Making Good Choices in the Social Sciences*. London: Sage.

The following two case studies can be found in the online resources. The first one should give you an idea of how to decide on a method for a research project and what this implies:

Minor, M., Smith, G. and Brashen, H. M. (2019) 'Cyberbullying in Higher Education: To Survey or not to Survey, that is the Question I ask of Thee', *SAGE Research Methods Cases*. doi: 10.4135/9781526490278.

The second case study demonstrates the role of conceptual planning of a design and study:

Walsh, L. (2017) 'The Challenges and Benefits of Developing a Rigorous Conceptual Framework in Your Research', *SAGE Research Methods Cases*. doi: 10.4135/9781473975026.

The following article, in the online resources as well, describes the process of quantitative research:

Vantieghem, W. and Van Houtte, M. (2018) 'Differences in Study Motivation within and between Genders: An Examination by Gender Typicality among Early Adolescents', *Youth & Society*, 50(3): 377–404.

Research Methodology Navigator

Orientation

- Why social research?
- Worldviews in social research
- Ethical issues in social research
- From research idea to research question

Planning and design

- Reading and reviewing the literature
- Steps in the research process
- Designing social research

You are here in your project

Method selection

- Deciding on your methods
- Triangulation and mixed methods

Working with data

- Using existing data
- Collecting new data
- Analyzing data

Reflection and writing

- What is good research? Evaluating your research project
- Writing up research and using results

9

TRIANGULATION AND MIXED METHODS

──How this chapter will help you──────────────────────────────

You will:

- understand the limits of using single approaches in social research
- appreciate the arguments for combining various procedures in a research project
- understand the concept of mixed methods
- understand the concept of triangulation, and
- see that combining methods can be productive, if you use them to collect or analyze data on different levels.

Combining Different Approaches

At a later point in this book, we will address the limits of specific methods and research approaches (see Chapter 13). In the development of social research, the idea of using several methods and approaches in the same study has attracted varying attention. Keywords here are triangulation and, more recently, mixed methods research. Common to these approaches is the use of multiple methods.

Relevance of using multiple methods in social research

An argument for using multiple methods in social research was advanced in the mid-twentieth century by Barton and Lazarsfeld (1955). The authors refer to the ability of qualitative research with small numbers of cases to make visible the relations, causes, effects and dynamics of social processes that cannot be found through statistical analyses of larger samples. According to their view, qualitative and quantitative research will be used at different stages of a research project. Qualitative research would mainly be used at the beginning, though it can also be employed subsequently for the interpretation and clarification of statistical analyses.

The debate about qualitative or quantitative research, which was originally informed by epistemological and philosophical standpoints (see Becker 1996, Bryman 1988 and Chapter 2 for overviews), has shifted towards issues of research practice concerning the appropriateness of each approach. Wilson states that in terms of the relation between the two methodological traditions, 'qualitative and quantitative approaches are complementary rather than competitive methods [and the] use of a particular method ... rather must be based on the nature of the actual research problem at hand' (1982, p. 501).

At the end of Chapter 8, the importance of deciding between qualitative and quantitative research was highlighted. For many research problems, however, a decision may lead to narrowing the perspective on the issue under study. The number of research problems that require a *combination* of qualitative and quantitative approaches, and thus of several perspectives on what is studied, is growing. Therefore, we will consider approaches for (a) combining methods within either qualitative or quantitative research and also for (b) combining qualitative and quantitative research within the same study. Bryman (1988, 1992) has identified 11 ways of combining quantitative and qualitative research, as follows:

1 The logic of triangulation means checking qualitative findings against quantitative results.
2 Qualitative research can support quantitative research.
3 Quantitative research can support qualitative research.
4 Integration may provide a more general picture of the issue under study.
5 Structural features are analyzed with quantitative methods and processual aspects with qualitative approaches.
6 The perspective of the researchers drives quantitative approaches, while qualitative research emphasizes the viewpoints of participants.
7 The problem of generality can be solved for qualitative research by adding quantitative findings.

8 Qualitative findings may facilitate the interpretation of relationships between variables in quantitative datasets.

9 The relationship between micro and macro levels in an area under study can be clarified by combining qualitative and quantitative research.

10 Using qualitative or quantitative research can be appropriate at different stages of the research process.

11 Hybrid forms may use qualitative research in quasi-experimental designs. (see Bryman 1992, pp. 59–61)

Overall, this classification includes a broad range of variants. Items 5, 6 and 7 rest on the idea that qualitative research captures different aspects to quantitative research. Theoretical considerations are not very prominent in the list of 11 variants Bryman identifies, as the focus is more on the pragmatics of research.

Mixed Methods

In recent years, a combination of multiple methods is often discussed by using the keyword of 'mixed methods'. Supporters of mixed methodologies are interested in combining qualitative and quantitative research pragmatically, seeking to end the 'paradigm wars' between the two approaches. Tashakkori and Teddlie (2003b, p. ix) declare this approach to be a 'third methodological movement'. They see quantitative research as the first movement; for them, qualitative research is the second movement, and mixed methods research is the third movement – one that resolves all of the conflicts and differences between the first and second movements. Figure 9.1 arranges the mixing of methods between the two poles of qualitative and quantitative methods, which shall be mixed. On a more general level, it is the combination of words and numbers, which is intended. On a more basic methodological level, this most often means the combination of interviews (analyzed by coding), on the one hand, and questionnaires (analyzed by statistics), on the other hand (see Chapters 11 and 12 for details).

Mixed methods designs

When using mixed methods in your own project, the issue of how to design a mixed methods study will be most interesting on a practical level. Creswell and Poth (2017) have distinguished three forms of mixed methods design, namely:

1   Phase design, in which qualitative and quantitative methods are applied separately, one after the other (no matter in what order). Such designs can include two or more phases.
2   Dominant/less-dominant design, which is mainly committed to one of the approaches and uses the other only marginally.
3   Mixed methodology design, which links the two approaches in all phases of the research process.

Figure 9.1   Mixed methods

For the mixed methodologies approach, Tashakkori and Teddlie (2010), Creswell and Poth (2017) and Creswell et al. (2003) have suggested a more elaborate version of design that combines qualitative and quantitative research. Creswell et al. (2003, p. 211) see mixed methods as a design in its own right in the social sciences, and use the following definition:

A *mixed methods study* involves the collection or analysis of both quantitative and/or qualitative data in a single study in which the data are collected concurrently or sequentially, are given a priority, and involve the integration of the data at one or more stages in the process of research. (2003, p. 212)

Overall, mixed methods have been used increasingly since the end of the 1990s to overcome the tensions between qualitative and quantitative research. Here, a rather pragmatic methodological approach is chosen.

One issue to keep in mind in this context is the differences in worldviews behind quantitative and qualitative research (see Chapter 2) when mixing them in one project: whereas the first is seeking to collect data seen as facts and independent of particular participants, the latter is interested in the subjective views and experiences of its particular participants. Given this difference in the epistemological status of an answer in a questionnaire and that in an interview, for example, we should simply equate the two statements. These differences are often neglected when the combination of the two approaches is reduced to applying two methods for producing two types of data. What is needed is a combination of approaches that also takes their differences into account on a theoretical (and epistemological) level. In this context, the concept of triangulation has a stronger relevance than proponents of mixed methods – as in the list of Bryman, mentioned earlier – would see for it.

Box 9.1   Research in the real world

A study of the study success and behavior of non-traditional students

Alheit, Rheinländer and Watermann (2008) did a study that combined a quantitative online survey (N = 886 participants; see Chapter 11) with 112 narrative interviews selected using theoretical sampling (see Chapter 7) of data from six German higher education institutions from two points in time (1998 and 2005; N = 2x around 400 survey respondents). The topic was the study success and study behavior of so-called non-traditional students. This group of students does not immediately begin to study at university after finishing school, or they have obtained their school qualifications via an alternative route and come from uneducated family backgrounds. The interesting aspect here is that the typologies found in analyzing the narrative interviews (see Chapter 11) were set in relation to the quantitative data and could be partially confirmed. The analytic techniques of maximal comparison known from analyzing interviews (see Chapter 12) were applied to the quantitative analysis, and distributional trends of the 'qualitative' type in the qualitative data could be found. The study is methodologically interesting for its mutually complementary findings and for its amendments to (e.g. quantitative) findings by integrating them with the other (e.g. qualitative) findings.

Triangulation

Although Bryman's list for using multiple methods (see above) files triangulation under the confirmation of (qualitative) results by using a second (quantitative) approach, a broader discussion about triangulation has developed over the years. A stronger focus on methodological issues around combining multiple methods is taken in these discussions. In the social sciences, triangulation means to view a research issue from at least two vantage points. Mostly, analysis from two or more points is realized by using multiple methodological approaches. As a strategy for grounding empirical research and its results (see Chapter 13), triangulation has attracted a lot of attention in the context of qualitative research. In particular, the conceptualization provided by Denzin has proved popular (1970/1989). Denzin initially understood triangulation as a strategy of validation (see Chapter 13), but after severe critiques (see Flick 1992 for an overview) he has developed a broader concept. As a result, Denzin has distinguished four forms of triangulation:

- *Triangulation of data* combines data drawn from different sources and at different times, in different places or from different people.
- *Investigator triangulation* is characterized by the use of different observers or interviewers to balance out the subjective influences of individuals.
- *Triangulation of theories* means 'approaching data with multiple perspectives and hypotheses in mind ... Various theoretical points of view could be placed side by side to assess their utility and power' (1970, p. 297).
- Denzin's central concept is *methodological triangulation* 'within method' (e.g. through using different subscales within a questionnaire – see Figure 9.2) and 'between method' (see Figure 9.3): 'To summarize, methodological triangulation involves a complex process of playing each method off against the other so as to maximize the validity of field efforts' (1970, p. 304).

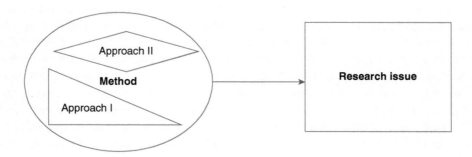

Figure 9.2   Within-methods triangulation

Figure 9.2 illustrates the idea of triangulating two methodological approaches in one method. For an example, see the episodic interview in Chapter 11, which triangulates asking questions with narratives in one form of interview.

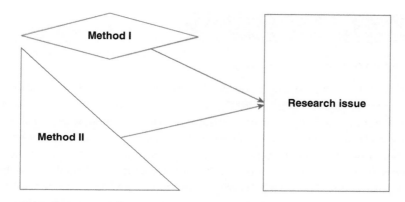

Figure 9.3 Between-methods triangulation

Figure 9.3 illustrates the concept of triangulating two independent methodological approaches, which means to combine methods such as interviewing and observation. It may also include the combination of a qualitative and a quantitative method.

Definition of triangulation

From these forms, we may develop a definition of triangulation: triangulation means that you take different perspectives on an issue in your study or in answering your research questions. These perspectives can be substantiated by using several methods or several theoretical approaches. Furthermore, triangulation may refer to combining different sorts of data against the background of the theoretical perspectives which you apply to the data. As far as possible, you should treat these perspectives on an equal footing. At the same time, triangulation (of different methods or of data sorts) should provide additional knowledge. For example, triangulation should produce knowledge on different levels, which means it goes beyond the knowledge made possible by one approach and thus contributes to promoting quality in research.

Triangulation within qualitative research

Triangulation can involve combining multiple qualitative approaches. The following case study illustrates the definition of triangulation provided above.

Box 9.2 Research in the real world

Triangulation within qualitative research

In Chapter 4, I mentioned our study on the situation of homeless adolescents with chronic illness (Flick and Röhnsch 2007, 2008). In this study, we combined three methodological approaches: (a) participant observations in the everyday lives of the adolescents; (b) episodic interviews with the adolescents about their concepts of health and illness, and their experiences within the health system; and (c) expert interviews with social workers and physicians about the healthcare situation of adolescents living in these circumstances. Thus, we achieved triangulation between methods and between different kinds of data (statements and observations). We also achieved within-method triangulation, since the episodic interviews combined two approaches ('questions' and 'narrative stimuli'), leading to two sorts of data ('answers' and 'narratives') (see Chapter 11). At the same time, different theoretical approaches were brought together: (a) an approach oriented towards interactions and practices in the group under observation; (b) an approach based more on narrative theories in interviews with the adolescents; and (c) an approach focused on expert knowledge as a specific form of knowledge in the expert interviews.

Triangulation within quantitative research

Similarly, you can use triangulation to combine several quantitative approaches. For example, you can use several subscales in a questionnaire, or combine several questionnaires, or complement questionnaires with standardized observations.

Triangulation of qualitative and quantitative research

Finally, triangulation may involve the combination of qualitative and quantitative research. Triangulation of qualitative and quantitative research often becomes concrete at the level of the results produced. It is on that level that Kelle and Erzberger (2004) focus in their discussion on combining the two approaches. They discuss three alternatives:

1   Results may *converge.* That is, the results confirm each other. They may also partly confirm each other and support the same conclusions. For example, statements from a representative survey with standardized questionnaires may align with statements from semi-structured interviews with a part of the sample in the survey.

2   Results may focus on different aspects of an issue (e.g. the subjective meanings of a specific illness and its social distribution in the population), but be *complementary* to each other and lead to a fuller picture. For example, interviews may provide results that complement (through, say, deepening, detailing, explaining or extending) findings obtained from questionnaire data.

3   Results may be *divergent or contradictory.* For example, in interviews, you receive different views from those provided by questionnaires. You could then take this as the starting point for further theoretical or empirical clarification of the divergence and the reasons behind it.

In all three alternatives, similar questions arise. For example, how far did you take into account the specific theoretical background of the two methods that you used in collecting and analyzing your data? Are any differences simply the result of the differing understanding of realities and issues in qualitative and quantitative approaches? Should convergences that go too far make you sceptical, rather than seeing them as a simple confirmation of the one by the other result? And, finally, how far are the two approaches and their results seen as equally relevant and independent findings, so that using the concept of triangulation is justified? How far is one of the approaches reduced to a subordinate role, such as merely making plausible the results of the other, more dominant, approach?

Procedures for triangulating qualitative and quantitative research

1   In the first step, you should consider how developed is the state of knowledge about your research issue and which empirical approaches are required, perhaps in combination.

2   To triangulate various approaches of social research, you will take the different theoretical positions and distinctions of the approaches into account when you plan your concrete methodological procedures.

3   In planning such a study, you can take the different designs suggested by Miles and Huberman (1994, p. 41) as an orientation, as illustrated in Figure 9.4. In the first design, you pursue both strategies in parallel. In the second design, one strategy (for example, continuous observation of the field) will provide a basis for it. You can use this basis to plan the waves of a survey. The several waves are related to

the observation from which these waves are derived and shaped. In the third combination, you will begin with a qualitative method, e.g. a semi-structured interview. These interviews are followed by a questionnaire study as an intermediate step, before you deepen and assess the results from both steps in a second qualitative phase. In the fourth design, you will do a complementary field study in order to add more depth to the results of a survey in the first step. The last step is an experimental intervention in the field to test the results of the first two steps.

4   The methodological approaches of collecting and analyzing data are based on the concepts of methodological and data triangulation formulated by Denzin (1989).

5   Finally, the presentation of results (see Chapter 14) reflects the combination of research approaches – by combining case studies with overviews in tables or charts, for example.

It is useful here to consider an example of research using triangulation in the way just considered (see Flick et al. 2012). Our next case study outlines this (see Box 9.3).

Overall, the approach of triangulation in social research can take us beyond some of the limitations of one-sided research. It uses for this goal the theoretical and methodological triangulation of several approaches in a complex design, making possible a more comprehensive understanding of the issue under study.

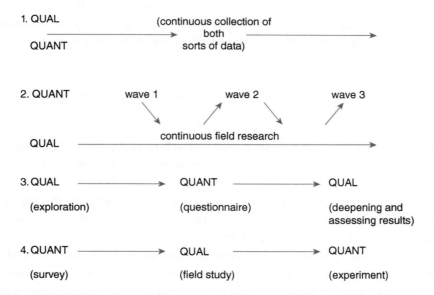

Figure 9.4   Research designs for linking qualitative and quantitative research

Source: Adapted from Miles and Huberman 1994, p. 41

Box 9.3 Research in the real world

Triangulation of qualitative and quantitative research

In a study about the issues of sleep and sleep disorders in institutional everyday life in nursing homes (see Flick et al. 2010, 2012; Garms-Homolová et al. 2010), we took different perspectives towards this problem in analyzing its relevance for routines of care and the development of disease. In a number of nursing facilities, we analyzed routine data in a longitudinal perspective. Repeated assessments of around 4000 nursing home residents were analyzed three times, each after a one-year gap. We looked for the interrelations between any sleep problems documented and the number and intensity of diseases during this period. Parallel to such secondary analyses, we conducted interviews with nurses and doctors in the same facilities, and complemented them with interviews of relatives of residents in this context. These interviews focused on the perception of the problem within care routines and how these problems were dealt with. The design was oriented on the second suggestion by Miles and Huberman (1994; see Figure 9.4). This study allowed for analysis of the issue on two levels: (a) the reality of care and (b) the professional perception of the problem. For this purpose, it combines two methodological perspectives: a quantitative secondary analysis of routine data, and a qualitative primary collection and analysis of interviews. On the methodological level, a triangulation of several approaches is realized. This research project is an example of integrated social research, not only because the results of qualitative and quantitative analyses refer to each other, but also because the two approaches are integrated within a comprehensive design.

Digital Research as a Complementary Strategy

Often, the Internet is used as a context for social research because of the limitations of traditional forms of social research. Some of these limitations are practical ones concerning costs and time. However, there are also more general limitations. For example, you can reach people over greater distances when you use a web survey instead of a paper survey. You will reach some people only via the Internet, for instance as members of a specific online community (see Chapter 11). At the same time, limitations of digital research have been identified. They include limited response rates, unclear sampling frames, the restriction of samples to people who use the Internet, and so on.

To overcome these limitations, several authors suggest digital and traditional research, such as combining paper and web surveys (see Fielding 2018). This can be done in two ways. Either you run your study in two phases, sending out a questionnaire by mail to a sample and sending a second electronically to another sample via the Internet, or you offer your respondents in a postal questionnaire survey the option to complete the questionnaire either on paper or online. In both cases, you combine the advantages of the two forms of research. Similarly, you could complement face-to-face interviews with online interviews (or focus groups). All in all, online research extends the options for which methods to mix or which perspectives to triangulate or integrate in your study.

Pragmatism and the Issue as Points of Reference

The preceding parts of this chapter have demonstrated that the rhetoric of strict separation (or incompatibility) between particular research approaches has its limits. The point of reference for choosing and combining research approaches should be the requirements of the issue under study, rather than single-sided methodological certainties and claims. Combinations can mean that you link qualitative and quantitative research as well as triangulating different qualitative or different quantitative approaches. At the same time, you should see the combination of research approaches pragmatically. That is, you can ask yourself: what is necessary for a sufficiently comprehensive understanding of the issues in your study? What is possible in the given circumstances of your own resources and in your field of study? When is it worth making the extra effort to combine methods? In making such a decision, consider whether the different methods will really focus on different aspects or levels or whether they will merely capture the same in very similar ways.

It is helpful here to differentiate between various possible levels of research, such as the four that follow, each of which is explained through examples.

Subjective meaning and social structure

To understand the subjective meaning of a phenomenon – for example, of a disease – patients are interviewed. This is complemented by analysis of the frequency and distribution of the disease in the population, or in subpopulations with different social backgrounds (see Flick 1998b).

State and process

The situation and the current practices of adolescents living on the street are analyzed for issues of health and illness by using participant observation as an approach.

This description of a state is combined with interviews revealing which processes in the particular adolescents' biographies have led them to their current situation (see Flick and Röhnsch 2007).

Knowledge and practice

Counselors' subjective theories of trust held in relation to clients are reconstructed in interviews. These are juxtaposed with conversation analyses of consultations done by these counselors with clients in order to find out how trust is built up or impeded (see Flick 1992).

Knowledge and routine

Nurses' knowledge about the effects of sleep problems on the health of nursing home residents is analyzed in interviews. In addition, routine data are analyzed which are based on diagnoses of certain diseases and documentations of sleep disorders (see Flick et al. 2010, 2012).

What You Need to Ask Yourself

The following question should help you to assess what the benefits and problems will be if you use (mix or triangulate) several methodological approaches in your study.

Box 9.4 What you need to ask yourself

Combining methods

1 What do you expect to gain from combining several methods?
2 Does this justify the extra effort in doing your research?
3 Will this be acceptable for your target group?
4 Does this fit with your resources, in particular time?
5 How do you think you will relate the various data (see Chapter 12 for more details)?

What You Need to Succeed

When deciding on using and combining several methods, you should consider the questions listed in Box 9.5. These questions can be used to inform the planning of your own study and also in assessing the studies of other researchers.

Box 9.5 What you need to succeed

Combining methods

1   What are the limits of the single method which can be overcome by combining several methods?
2   What is the extra gain in knowledge that you can expect from combining methods?
3   Do the methods really address different levels or qualities in the data, so as to justify their combination?
4   Can the extra effort required in combining methods be accommodated in your framework of research (resources, time, etc.)?
5   Are these efforts proportionate to the gain in knowledge they make possible?
6   Are the combined methods compatible with each other?
7   How should you sequence the methods you will use? How will sequencing affect the study?
8   How far are the methods applied according to their characteristics, so that their specific strengths are taken into account?

What you have learned

- Every method has its limitations in what it can grasp and how.
- Distinct limits can be identified for quantitative research and for qualitative research.
- Combinations of research strategies can help to overcome such limits.
- Triangulation and mixed methods provide ways of combining methods.
- Combinations should be grounded in the issue under study and in the additional gain of knowledge they make possible.

What's next?

This first book is the updated edition of the basic text that introduced the concept of triangulation to qualitative research in a general way. Reading it will give you more ideas about the background to and developments of the discussion about the concept:

Denzin, N.K. (1989) *The Research Act: A Theoretical Introduction to Sociological Methods*, 3rd edn. Englewood Cliffs, NJ: Prentice Hall.

In this short book, triangulation is introduced in more detail and in relation to the discussion about mixed methods:

Flick, U. (2018) *Doing Triangulation and Mixed Methods*. London: Sage.

The following book chapter discusses the strategy of triangulation in qualitative research in more detail:

Flick, U. (2018) 'Triangulation', in N.K. Denzin and Y.S. Lincoln (eds) *The SAGE Handbook of Qualitative Research*, 5th edn. London: Sage. pp. 444–61.

This next book outlines mixed methods research:

Greene, J.C. (2007) *Mixed Methods in Social Inquiry*. San Francisco: Jossey-Bass.

This next two texts apply the ideas of triangulation and mixed methods in different areas. The following case study can be found in the online resources. It should give you an idea of how a relevant problem is analyzed using a mixed methods approach:

Edwards, K. and Dardis, C. (2014) 'Conducting Mixed-Methodological Dating Violence Research: Integrating Quantitative Survey and Qualitative Data', *SAGE Research Methods Cases*. doi: 10.4135/9781446273050135 16582.

This article, also in the online resources, presents research linking qualitative and quantitative research, framing it in the concept of triangulation:

Flick, U., Garms-Homolová, V., Herrmann, W.J., Kuck, J. and Röhnsch, G. (2012) '"I can't prescribe something just because someone asks for it ...": Using Mixed Methods in the Framework of Triangulation', *Journal of Mixed Methods Research*, 6(2): 97–110.

WORKING WITH DATA

The central part of a research project consists of working with data either using available data or collecting new ones and analyzing them. This fourth part of the book explains these processes.

Chapter 10 addresses ways of using existing data in real world and virtual contexts. This kind of research becomes more relevant when students and researchers do not have the resources (time, money, etc.) for collecting their own data. This approach can also become relevant, if the research question suggests using existing data – in the context of analyzing documents or for a secondary analysis of data collected by other researchers or in an earlier project. A third reason for doing so is that data for answering a research question already exist and that it would (a) be a waste of researchers' resources and (b) unethical to collect data given that already existing data are available.

Chapter 11 introduces three main aspects of collecting one's own data. The first is the use of questionnaires or interviews; the second is the use of observation; and the third is how to obtain and document information in research.

Chapter 12 focuses on three major forms of data analysis. The first is content analysis; the second is descriptive statistics; and the third is interpretative analysis for qualitative data. Case studies are discussed at the end.

Research Methodology Navigator

Orientation

- Why social research?
- Worldviews in social research
- Ethical issues in social research
- From research idea to research question

Planning and design

- Reading and reviewing the literature
- Steps in the research process
- Designing social research

Method selection

- Deciding on your methods
- Triangulation and mixed methods

You are here in your project

Working with data

- Using existing data
- Collecting new data
- Analyzing data

Reflection and writing

- What is good research? Evaluating your research project
- Writing up research and using results

10

USING EXISTING DATA

─How this chapter will help you─────────────────────────

You will:

- see the advantages of using existing data for your research
- understand the difference between information and data
- know about methods of secondary data analysis in social research
- understand what you can do with data from the Internet
- have an idea about using social media and Big Data for your research and what they can provide, and
- know the limits of using secondary data.

Before we address methods for collecting new data for your own research, in this chapter, we will consider the use of existing data. You may have problems of time and other resources in your own study, which might suggest refraining from collecting new data and looking for alternatives. I will discuss reanalyzing other researchers' data, using real-world and virtual documents, and the Internet and social media as resources providing information that might be turned into data. In the last step, we will add some considerations of the phenomenon of Big Data. However, using existing data can be an alternative to collecting your own data – in both cases, though, similar procedures for analyzing the data (see Chapter 12) you work with will be the next step (see Figure 10.1).

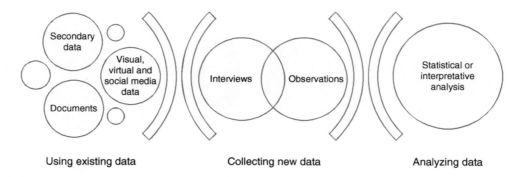

Figure 10.1 Using, collecting and analyzing data

What Does Existing Data Mean?

Existing data are not the result of the researcher's application of data collection methods such as interviews. Rather, the researcher uses data collected before, by other researchers, or in routine procedures of monitoring, or documents that exist independently of the current study – for example, administrative records or diaries written by people in their everyday lives. Promising sources for existing data are the Internet and social media, and increasingly the so-called 'Big Data'.

Why Use Existing Data?

In several contexts, interest in existing data is growing for a number of reasons. One is limited resources, which prevent researchers from collecting their own data, a lack of money, a lack of people to do the data collection and, mainly, a shortage of time. This may all equally apply to research in the context of obtaining a university degree, such as an empirical master's thesis. But there are other reasons, such as the fact that there are so many sets of data available for research already. As many funders require researchers to make their project data available in an archive, it may be reasonable first to search for projects very similar to your own planned study and to find out whether those data are available, can be obtained for your own use, can be re-analyzed, and so on. Furthermore, most modern states collect a wide range of data for administrative purposes: official statistics, evaluative data within education and health systems, regular monitoring of social processes and developments, or for security purposes, such as CCTV. Again, it can be helpful to check first whether any of these data fit your own project and study purpose, whether these data are available or accessible, and if they really fit the study and its purpose. Finally, it might be seen as an ethical problem to collect new data if already existing data could be used in an equal way.

Information and Data

The Internet can be used as a source for finding data to re-analyze, but it is necessary first to reflect on the relationship between information and data: 'The Internet and mobile apps present large opportunities for collecting multifarious and real-time information *from* citizens and *on* citizens, their behaviour and their interactions' (Corti and Fielding 2016, p. 1). The Internet offers more or less unlimited amounts of information about everything and anything. This information can be attractive for doing research, but it is not immediately what we understand as data. Even if the information you find is based on data and analyses by other researchers in earlier studies, there are some steps to take before this can become data for your own study. These steps include the following:

1   Select material (according to the relevant criteria or procedures – see Chapter 7) for your study.
2   Define what level of quality these data should have to be suitable for your study (see also Chapter 13).
3   Decide on methods for analysis of the data you use (see Chapters 8 and 12) and whether or not the information you find fits somewhere within this kind of analysis.
4   Define ways for securing this information as data – can you save this information to your computer hard drive or print it out in order to avoid it disappearing during your analysis because someone decides to take it offline or fundamentally rework it?

By taking only these steps into account, you may turn perhaps arbitrary, volatile and unreflected on information into documented, transparent, assessable and systematic data for your study. Further, when you take these steps into account, you can save the time and efforts of collecting data if you use existing data. There are basically four ways to use existing data:

1   Do a secondary analysis of earlier research or data, for example that found in archives or published
2   Work with documents to be found in archives or on the Internet
3   Analyze social media, and
4   Address Big Data.

We will address these four ways in turn in this chapter.

Secondary Analysis

The term 'secondary analysis' means that you are analyzing data that were not collected for your own research project. Instead, you use existing datasets that were produced for other purposes. Hakim's (1982, p. 1) definition of secondary analysis as 'any further analysis of an existing dataset which presents interpretations, conclusions or knowledge additional to, or different from, those presented in the first report on the inquiry as a whole and its main results' has been an orientation in the literature on this field (e.g. Dale et al. 1988). Secondary analysis is most often a quantitative re-analysis of quantitative data, but it has been widely discussed and used for qualitative data and analysis as well (see Corti 2018; Largan and Morris 2019). Secondary analysis may refer to re-analyzing data within a single data set (e.g. Fielding and Fielding 2000), to revisiting one classical study with new questions (Savage 2005), or to reanalyzing data across several data sets (Bishop and Kuula-Luumi 2017). It can focus on quantitative or qualitative data alone or refer the latter to quantitative evidence of the issue under study as well (Irwin et al. 2012). Secondary analysis is the analysis of data by researchers who have not been involved in data collection or the study itself. The assumption often is that the data for reanalysis are high quality data – based on rigorous sampling procedures, often with a wide variety of participants in their regional back-ground, for example, if official statistical data are reanalyzed. Quantitative secondary analysis allows for longitudinal analysis as the data are repeatedly collected and updated in several waves of data collection, for example in monitoring educational or health-related practices. Secondary analysis allows for a subgroup analysis – for example, to concentrate on specific age groups in a representative survey sample – and it can allow for cross-cultural analysis if data from EU surveys are used (see Bryman 2016, pp. 311–12).

Kinds of secondary analyses

Here, we may distinguish between re-analyzing data from other research projects and data produced for purposes other than research. In the second category, we find data from federal bureaux of statistics, which are collected, elaborated on and analyzed for monitoring purposes. They can be used for a number of research questions. In the first category, you will find data produced and analyzed by other researchers in studying a specific issue that now may be used again by other researchers for their own research questions. This can be organized through the direct cooperation of researchers with other researchers. But there are also several institutions that collect datasets from research pro-jects and provide them for other researchers, who pay for the right to use these data. Examples of such institutions are GESIS in Germany and GALLUP in the USA. In several contexts, datasets are developed as 'public-use files' and made available to interested researchers for further work with them. Corti and Fielding (2016, pp. 2–3) distinguish online data sources or open data sources, such as the US government's Data.gov portal, from commercial data sources, such as the UK Collaborative Online Social Media Observatory. These kinds of sources can be used, but because of the mass of data and the restrictive terms of use, they are not necessarily easy to use. Finally, Corti and Fielding mention dedicated academic research online portals, such as the official network of social sciences data services provided by the Council of European Social Service Data Archives (CESSDA). Beyond that, they discuss 'other kinds of data repositories, spanning academic libraries, publisher-related repositories, and dedicated commercial research data storage services' (Corti and Fielding 2016, p. 3). Corti and Wathan (2017) describe the standards of providing data sets for secondary analysis in Box 10.1.

Box 10.1 Research in the real world

Collecting secondary data sets

'A structured metadata record is created that captures core descriptive attributes of the study and resulting data. The DDI metadata standard is used by many social science archives across the world. The DDI is a rich and detailed metadata standard for social, behavioral and economic sciences data used by most social science data archives in the world. A typical DDI record will contain mandatory and optional metadata elements relating to study, data file and variable description:

- Study description elements contain information about the context of the data collection, scope of the study (e.g. topics, geography, time, data collection methods, sampling and processing), access information, information on accompanying materials and provides a citation;
- File description elements indicate data format, file type, file structure, missing data, weighting variables and software used;
- Variable-level descriptions set out the variable labels and codes, and question text where available.' (Corti and Wathan, 2017, p. 495)

Types of data for secondary analysis

In the context of social work research, Radey (2010) discusses the various types of secondary data one might find for a secondary analysis:

- Cross-sectional data are data collected at a specific point in time.
- Time-series data measure a phenomenon over time by asking varying respondents the same questions over time.

- Longitudinal data return to the same people over time.
- Survey data are collected in different forms for different purposes in questionnaires.
- Official statistics are statistical data gathered and generated not for research but as a result of another goal.
- Official records are data gathered and generated for non-research goals by social service agencies, for example. (Radey 2010, p. 166)

To do your study with secondary data, you should think about which type of data is most appropriate for your purpose, select a data set and familiarize yourself with the methodology that the original data collection was based on.

Figure 10.2 arranges the sources for using existing data from the area of research to the world of official documents and the realm of everyday knowledge up to the ubiquitous availability of the Internet as a source.

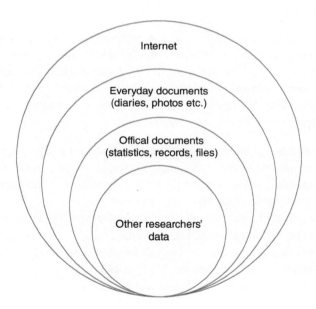

Figure 10.2   Sources for using existing data

Steps for secondary analysis of quantitative data

Radey also outlines the major stages of a secondary analysis of (quantitative) data:

- problem definition and question formulation

- selection of a data set
- selection of sample and measures
- assessing data reliability and handling error. (2010, p. 168)

Selecting and weighting secondary data

When you use secondary data, you should consider some questions and problems arising from this. First, you should check whether these data fit your research question: do they include the information necessary to answer it? Second, you should assess whether the form of elaboration in which the data are available corresponds to the aims of your study. A simple example will illustrate this: if your research question refers to a distribution of age in one-year steps but the available dataset has classified the studied people into age groups (e.g. in five-year steps), and if this classification cannot be traced back to the single value in years, this might produce problems in the usability of the data for your purpose. A second example: many epidemiological studies have been based on death-cause statistics within local health authorities for reconstructing the frequencies of certain diseases as causes of death. Here, in many cases, it was not the single death certificate which was used as data but the classification done by the particular health authority. Here again, the resulting problem is that the classification of different causes of death (or disease) into categories will not necessarily be the same as the researchers need or would apply to their study and research question. Mistakes made in allocating a particular case to a category cannot be assessed in such a dataset (see Chapter 12). The more general problem is that the data are available in aggregated form only, i.e. elaborated on and processed, and the original raw data are not accessible.

Archives in the face of the Internet

Gill and Elder (2012) discuss how the ideas of the archive and data have changed in the face of the Internet. They see three aspects to this: (1) the *traditional archive*, which is still based on storing paper versions of material, which has to be accessed on site, and which continues to exist unaffected by digital developments; (2) the *archive online*, which means that the traditional paper records are *also* available online and become more easily accessible as a main development; and (3) the *online archive*, which refers to seeing the Internet as a big archive in which, for example, traces of online communication about an issue are stored and accessible for research. It can then be seen as a site of doing research about a topic and a subject in itself – for example, when the way of communicating online becomes the subject of a study. Each of the three versions of archives can be fruitful for secondary analysis of their contents but vary in their accessibility to researchers.

Problems of secondary analysis

Several limitations of secondary analysis have been repeatedly discussed. First, the secondary researchers' lack of familiarity with the data may cause problems because the secondary analyst was not part of the data collection and elaboration of the original data. A second issue is that available datasets can be very complex and secondary analysts must reduce this complexity, for example by identifying the relevant parts and variables in the original data. The available data may not be the original raw data but already processed data, which means that secondary analysts will have limited or filtered access to what was said by respondents. Third, researchers will have limited or no control over data quality compared to collecting their own data. And finally, in the existing data, key variables or aspects necessary for one's own analysis may be absent as they were not in the focus of the original research and its research questions.

Working with Documents

The approaches presented so far in this chapter have in common that you are *reusing* the data of other researchers – data such as answers in a survey, a narrative in an interview or observations and descriptions in the field or laboratory. As an alternative, you may use materials already existing and not produced by (other) researchers, such as documents resulting from an institutional process. These may be texts or images, which can be analyzed in a qualitative or quantitative way depending on the research question. The analysis of documents can refer to existing materials – like diaries – which have not yet been used as data in other contexts. Sometimes they refer to existing datasets from other contexts – like official statistics, which have been produced not for research but for purposes of documentation. A definition of document analysis with a focus on qualitative research is given by Bowen: 'Document analysis is a systematic procedure for reviewing or evaluating documents – both printed and electronic (computer-based and Internet-transmitted) material. Like other analytical methods in qualitative research, document analysis requires that data be examined and interpreted in order to elicit meaning, gain understanding, and develop empirical knowledge (…). Documents contain text (words) and images that have been recorded without a researcher's intervention' (2009, p. 27).

Qualitative and quantitative analysis of documents

Although the above definition has a focus on qualitative analysis of documents it is also instructive for quantitative approaches to documents. Tight (2019) shows that the major quantitative approaches discussed here in Chapter 12, such as content and statistical analysis, are applied to existing documents as well.

Alternatively, you may use existing documents for a qualitative analysis. Wolff (2004, p. 284) suggests as a definition of what is generally understood as 'documents': They are 'standardized artifacts, in so far as they typically occur in particular formats: as notes, case reports, contracts, drafts, death certificates, remarks, diaries, statistics, annual reports, certificates, judgments, letters or expert opinions'.

Again, you can use documents produced for your study. For example, you might ask a group of people to write a diary over the next 12 months and then analyze and compare what was noted in these diaries. Or you might use existing documents, for example the diaries written by a specific group of people (e.g. patients with a specific diagnosis) in their everyday lives independent of the research. Documents are mostly available as texts (in printed form) and more and more also or exclusively in electronic form (e.g. in a database).

Although visual materials can be seen as documents as well, the following definition shows that the analysis of documents mostly means written documents:

> What can be included as a 'document' in social research covers a potentially
> broad spectrum of materials, both textual and otherwise. There are of course
> 'official' records of various kinds – organisational and 'state' documents designed
> as records of action and activity such as large data sets and public records. There
> are also everyday documents of organisations and lives – notes, memoranda, case
> records, email threads and so forth; semi-public or routine documents that are
> at the heart of everyday social practice. There are also private papers of various
> kinds that we can treat as documentary data or evidence – for example diaries,
> testimony, letters and greetings cards. (Coffey 2014, p. 367)

Here, the guiding idea is that most parts of our daily life, and even more of institutional life, leave traces somewhere as they are documented. Again, a driving force behind analyzing these traces, instead of talking to people in interviews (for example), is the assumption that the interesting events become available without the filters of individual memory and meaning making. But again, documents are not 'just there' – they were made by someone for some purpose and become relevant for the research only through the researchers' interpretations. In particular, the intentions and purposes for documenting something in a specific form have to be taken into account in analyzing the documents. For documents as well, central methodological issues are what to select as material, what to select in the material and for what this selection is seen as relevant.

Many official or private documents are meant only for a limited circle of recipients who are authorized to access these documents or who are their subjects. Official documents allow conclusions about what their authors or the institutions they represent do or intend, or how they evaluate. Documents are produced for a certain purpose – such as to substantiate a decision or to convince a person or an authority. But that also means that documents represent issues only in a limited way. When analyzing them for research purposes, you should always consider who has produced a document, for whom and for what purpose. How documents are designed is a part of their meaning; how something is presented influences which effects are produced by a document.

From a practical point of view, the first step is to identify the relevant documents. To analyze official records, you have to find out where they are stored and whether or not they are accessible for research purposes. Then you have to make the appropriate selection: which of the existing records you will concretely use and why (see also Rapley 2018). In assessing the quality of documents, Scott (1990) proposes four criteria, which you can use for deciding whether or not to employ a specific document (or set of documents) for your research:

- Authenticity – is the evidence genuine and of unquestionable origin?
- Credibility – is the evidence free from error and distortion?
- Representativeness – is the evidence typical of its kind, and, if not, is the extent of its untypicality known?
- Meaning – is the evidence clear and comprehensible? (1990, p. 6)

The first criterion addresses the question of whether the document is a primary or secondary document: is it the original report of an accident, for instance, or is it a summary of this original report by someone who did not witness the accident itself? What was omitted or misinterpreted in writing this summary? Tertiary documents are sources for finding other documents, like the library catalogue lists primary source documents. Looking at internal inconsistencies or comparing with other documents, by looking at errors and by checking whether different versions of the same document exist, can assess *authenticity*. *Credibility* refers to the accuracy of the documentation, the reliability of the producer of the document, the freedom from errors. *Representativeness* is linked to typicality. It may be helpful to know of a specific record whether it is a typical record (which contains the information an average record contains). However, it can also be a good starting point if you know a specific document is untypical and to ask yourself what that means for your research question. In terms of *meaning*, we can distinguish between (a) the intended meaning for the author of the document, (b) the meaning for the reader of it (or for the different readers who are confronted with it), and (c) the social meaning for someone who is the object of that document. For example, the protocol of interrogation was written by the author in order to demonstrate that this was a formally correct interrogation. For a judge in court, the meaning of the content of the protocol is to have a basis for reaching a judgment. For an accused man, the meaning of the content of the protocol could be that he now has a conviction, which will have consequences for the rest of his life, when he tries to find a job, and so on. And for the researcher, the meaning of this protocol might be that it demonstrates how guilt is constructed in a criminal trial.

Box 10.2   Research in the real world

A student's research study on stigma and mental illness in political debates

In his research for obtaining his master's degree in education, Benedict Lechner (2019) analyzed documents from the parliamentary debates of a 'mentally-ill-help-law' in Bavaria, Germany. The data basis consisted of short hand documented, transcribed and published documents of plenary debates in the Bavarian parliament from April to June 2018, which could be found in the Internet. In a discourse analysis of the material (see Chapter 12), he could identify two figures of discourse. The first defines conflicting relationships between de-stigmatization of mental illness and the accusation of stigmatisation as a process and in the relation to the mentally ill person. In the second figure of discourse, several thematic entanglements of mental illness with help and care, criminalisation and punishment, security issues and stigmatisation of the mental ill people were dominant.

Visual Data as Documents: Photo and Film

Visual data, such as photos, film and video, have attracted increasing attention as documents that can be used as data in research (see Knoblauch et al. 2014, 2018). Again, we may distinguish two approaches: cameras can be used as instruments for data collection and images can be produced for research purposes; or existing images can be selected for the research and analyzed. One approach could be to analyze the photos in family albums and the family's or the individual's history as documented in these photos over time. In research with families or institutions, the analysis of self-presentations in photos or images of the members displayed on the walls of rooms can give insights into the structures of the social field (see Banks 2018). Media analyses also address films or TV series to analyze the presentation of certain problems and how a society deals with them (see Mikos 2014, 2018, and Denzin 2004 for this).

When you use existing material (texts, images, datasets) for your study, you will save time at the data collection stage, since this is limited to selecting from the existing material. But you should not underestimate the obstacles arising from the use of such material: photos, texts and statistics have their own structure that is often strongly influenced by who produced them and for what purpose. If this structure is not compatible with the demands for data arising from your own research question, it may produce selectivity problems. Solving these problems can sometimes be more expensive and time-consuming than collecting your own data. Thus, the use of secondary data in particular, and existing material in general, is only suggested if the research question gives you good reason.

Analyzing Internet Documents and Interactions

Bergmann and Meier (2004) go one step further beyond focusing on the content of virtual documents. Starting from a conversation analytic background, they suggest analyzing the formal parts of interaction on the Internet. Conversation analysis is interested in the linguistic and interactive tools (like taking turns, repairing, opening up closings; see Chapter 12) that people use when they communicate about an issue. In a similar way, the authors suggest that you identify the traces that online communication leaves for understanding how communication is practically produced on the web. Thus, Bergmann and Meier use electronic process data, that is 'all data that are generated in the course of computer-assisted communication processes and work activities – either automatically or on the basis of adjustments by the user' (2004, p. 244). These data are not simply ready to hand; rather, they must be reconstructed on the basis of a detailed and ongoing documentation of what is happening on the screen (and, if possible, in front of it) when someone sends an e-mail, for example. This includes the comments of the sender while typing an e-mail, paralinguistic aspects like laughing, and so on. It is important to document the temporal structure of computer-mediated communication. Here, you can use special software (like Lotus ScreenCam) which allows you to film what is happening on the computer screen, together with video recording the interaction occurring in front of the screen, for example.

Similarly, you may take websites as a medium of online interaction and analyze them for their content and for the means that are used for communicating this content. You may use qualitative methods for such a study – like a hermeneutic approach or a qualitative content analysis (see Chapter 12) – or approach these objects with quantitative methods like content analysis, or by analyzing the frequencies with which they are addressed or used. A specific feature of websites is the intertextuality of documents on the web, organized and symbolized by (electronic) links from one text (or one page) to other texts. This kind of cross-referencing goes beyond the traditional definition and boundaries of a text and links a large number of single pages (or texts) to one big (sometimes endless) text. Many websites are constantly updated; they change, disappear and reappear on the web. It is necessary, therefore, to always mention the date you last accessed a page when referring to it as a source. As with other forms of analyzing documents as means of interaction, you should ask: who produced this website, for whom and with what intentions? Which means are used to reach these goals?

The example of a case study of an ethnography of online communication in Box 10.3 illustrates this.

Box 10.3 Research in the real world

Online ethnography of communication

In her study, Hart (2014) analyzed the communication of Eloqi, which is a pseudonym for a for-profit start-up company that connected English language learners in China with trainers in the US. In her article, she describes the technologies she used to carry out her qualitative study, including online participant observations, online and offline interviews, qualitative coding, and qualitative data analysis. Hart collected 'examples of real communication between Eloqi's members (admins, trainers and students) to see that real communication in its larger context'. Her data were documented communications complemented by Skype calls recorded by software for documenting them (Audio Hijack Pro). The trainer–student communications were documented in audio recordings, which could be downloaded and analyzed. Hart ends her article with recommendations for doing similar studies – both on the level of technology and on that of methodology.

Social Media for Doing Research

It seems that social media have been 'discovered' for doing research now. As several authors show, defining in a clear way what social media are seems difficult. McCay-Peet and Quan-Haase (2017) mention a lack of formal definitions and refer to Papacharissi, who mentions that 'our understanding of social media is temporally, spatially, and technically sensitive.... How we have defined social media in societies has changed and will continue to change' (Papacharissi 2015, p. 1). The authors have surveyed suggestions for defining social media and summarize the results in their own version:

> Social media are web-based services that allow individuals, communities, and organizations to collaborate, connect, interact, and build community by enabling them to create, co-create, modify, share, and engage with user-generated content that is easily accessible. (McCay-Peet and Quan-Haase 2017, p. 17)

We can find a wide range of issues studied in the context of Facebook, Twitter, blogs and the like, as the following list of example research questions shows (see also Flick 2018a, Chapter 22).

Basic questions for planning social media research

Although social media offer much information at hand to use in research, research is more than 'just do it'. We need to apply a specific kind of data collection when using data from social media. Approaches to social media can be qualitative but most are quantitatively oriented. In their article 'Think before you collect', Mayr and Weller (2017, p. 110) suggest four basic questions to ask before carrying out any data collection from social media:

1   'Which social media would be most relevant for my research question(s)? (Single platform vs. multi-platform approach)'. This refers to the question of whether it will be enough to collect just tweets from Twitter, for example, or whether you should also have other social media activities included in your study, such as Facebook, if you want to understand politicians' use of social media in election campaigns.
2   'What are my main criteria for selecting data from this platform? (Basic approaches for collecting data from social media)'. Some platforms are more likely to be used by specific target groups and thus it may be more relevant to analyze blogs instead of tweets.
3   'How much data do I need? (Big versus small data)'. This is a question of sampling, but also of whether I should use just the published tweets (small data) or also reconstruct the traces they leave, and the reactions and comments they provoke, which may be beyond the tweeter's perception (Big Data – see below).
4   'What is (unproportionally) excluded if I collect data in this way (collection bias)?' This is a question of any distortion and bias I produce by concentrating on one approach, for example. Possible biases come from sampling, from access restrictions and from biased social media populations (2017, p. 114).

These questions refer to planning the use of social media data. They show that we are talking about research and designing research (see Chapter 7) here as well, even if we are not using the methods of data collection discussed in the next chapter. The following questions refer to what we might want to study by using social media as a database (see Flick 2018a, Chapter 22).

Example research questions and themes referring to social media

1   Presentation of self- and reputation management – e.g. how do users present themselves?
2   Action and participation of social media users – e.g. what are they doing with Twitter and with whom?
3   Uses and gratification – what are the benefits of using Instagram and for whom?
4   What are the positive and negative experiences of being involved in social media?
5   Usage and activity counts
6   Issues around the social context of users' social networks
7   The platform characteristics of Facebook, for example, and their impact on use and users.

Box 10.4 Research in the real world

Using Twitter as a research tool

Stewart localizes her research in the context of 'traditional ethnographic methods adapted for a geographically distributed digital-communications-based study' (2017, p. 254). Her interest was in the academic use of Twitter by her participants. First, the focus was on using social media to invite potential participants with a call for volunteers distributed on several social network sites, mainly Twitter. The latter is described by Stewart as a 'valuable platform for research into decentralized non-gate kept professional cultures' (2017, p. 254). From the response to her call, she selected 14 participants from the US, Canada, Europe, Latin America, Africa and Australia, and 'eight "exemplar" identities' (p. 256) who were ready to have their Twitter profiles assessed by the participants. Her participants also agreed to be openly identified in the research by their public Twitter use. Methodologically, Stewart applied participant observation, for which she created a specific Twitter account for observation, from which she observed several times per day how 'participants presented themselves, engaged with others and shared their work and that of other people' (2017, p. 257). Stewart tweeted in a limited way but used Twitter as a performative space. Participants were, in addition,

(Continued)

asked to select a 'representative 24-hour period' of their network activities and inform the researcher when this period was going on or had just finished. In studying these 24-hour reflections, Stewart extended her focus to other platforms on which the participants were active (such as Instagram). The participants made short reflective documents with screen captures of their activities over that period. Furthermore, the study includes 'profile assessments' of the eight exemplary identities concerning the assumed influence of the persons described in these identities. Finally, Stewart did 10 Skype interviews with participants (not on Twitter) during the process of participant observation after a few weeks. The transcripts of the interviews were collated with the reflective documents provided by the participants and analyzed by coding.

Problems with Using Digital Data

This chapter first focused on the secondary analysis of existing data and documents. In both contexts, the localization of data was increasingly linked to online and digital sources. While this facilitates the access and use of such data, it also includes risks and problems, which can be summarized in the question of how far we can trust digital data sources. Discussing the question of trust here is less about focusing on a general scepticism against the Internet or digitalization in general, and more about the technical aspects that may put that trust into question. Corti and Fielding (2016, p. 1) discuss a number of problems in the context of (re-)using digital qualitative data: 'Unknown provenance, lack of insight about sampling, error and bias, and broken links require us to rethink our trust in data and follow new best practice.'

Trustworthiness of online data sources: Sustainability and persistence

Corti and Fielding (2016, p. 2) outline an approach of 'FAIR' data publishing principles. These 'embrace the principles of **F**indability, **A**ccessibility, **I**nteroperability, and **R**eusability and focus on the specification of minimally required standard protocols, lightweight interfaces, and formats'. The main issue here is how far researchers can rely on the data they want to reuse, and on the quality of them. The solution for Corti and Fielding is to have the necessary information about the data and their origin, for example:

> To assess whether we can trust data, we need detailed information: about the
> provenance of the data to hand such as the reason for collecting the data, and

the sampling and methods used; about the content of the data, such as its shape, format, volume, and topics; and whether it can be used legally and ethically. (2016, p. 4)

Again, this concept of trust based on information shows that working with existing data from digital or other sources is more than 'just-do-it' research but is based on reflection, planning, creating a dataset (from other resources) and assessing the quality of the data and of what is possible with them for one's own study.

Big Data

The term Big Data refers to the trend of bringing all sorts of data and social media fragments together and analyzing them. Because of the enormous capacity of modern computers, the idea has developed that we need not waste time by thinking theoretically about planning sampling or other forms of research design, but should just calculate and process the data. Data were initially characterized by huge volume, high velocity and diverse variety, whereas now these features have been refined. Big Data are seen now as being 'exhaustive in scope, striving to capture entire populations or systems (n = all), fine-grained in resolution aiming to be as detailed as possible, and indexical in identification, relational in nature ... and flexible' (Kitchin 2017, p. 27). The danger connected to this kind of data and the discussions they stimulate is that they might seem to make obsolete other forms of research that do not work with Big Data. Kitchin critically discusses suggestions that the end of theory has arrived because the available data allow so much correlation that more theoretically informed research seems outdated and unnecessary. Arguments against this idea include the notion that data are not simply data (see the discussion about 'raw data as oxymoron' in Gitelman 2013). Kitchin discusses issues such as the 'hundreds of thousands of fake Twitter accounts seeking to influence trending and clickstream trails' (2017, p. 35). Big Data has commonly been approached with *small-scale* content analysis – looking at small numbers of users – or larger-scale *random or purposive samples* of tweets. Tinati et al. (2014, p. 665) demonstrate a new approach of displaying Twitter activities after an event based on a social network analysis approach (an example of such an event is the protests against the rise in university tuition fees in England in 2011 – see also Bryman 2016, p. 326). Their tool 'enables the metrics, dynamics and content of Twitter information flows and network formation to be explored in real-time or via historic data' (p. 668). They used it to analyze the role of Twitter in political activism in their example of protests against university tuition fees. They included 12,831 tweets from 4737 Twitter users from 8 October to 21 November 2011. 'They also show that there are individuals who, though not generators of content themselves, play a significant role in the flow of information by being significant re-tweeters' (Bryman 2016, p. 326). In another example, Procter et al. (2013) did a Twitter analysis following the August 2011 riots in England. In the appendix to their article, they listed their coding frame for tweet types, which is most informative.

What You Need to Ask Yourself

If you plan to work with existing data, you should be able to answer the questions in Box 10.5.

Box 10.5 What you need to ask yourself

Existing data

1   Why did I choose these specific kinds of documents?
2   What makes them relevant as data and not just information?
3   Who produced them and for which purpose?
4   How can I define the quality of these materials?
5   How did I select these materials (and not other ones also available)?
6   How well do they fit my research question and aims?
7   If they come from other research, how far do the data and how they were produced harmonize with my own research?
8   How should these materials be analyzed (see Chapter 12)

What You Need to Succeed

If you are conducting an empirical project, it will help to consider the questions and points in Box 10.6 around the use of existing data. These points can be borne in mind when doing your own study or when assessing the research of other researchers.

Box 10.6 What you need to succeed

Existing data

1   Be aware of any biases referring to research in the data resulting from the way and context in which they were produced.

2 Do the data you want to re-use, or re-analyze, or the documents you want to analyze provide the information necessary to analyze them in a comprehensive way and a way appropriate to your research question?
3 Check for any problems and limits in the accessibility of documents and data you want to use.
4 Don't confuse simple information you find on the Internet with data you want to use.
5 Be aware that you must turn this information into data which can be used in a similar way to data you would have collected yourself.

What you have learned

- Reusing other researchers' data has advantages in saving time and resources but can sometimes be tricky if the data do not perfectly fit your own research question.
- You need to know a lot about the background of other researchers' data to be able to use them.
- Documents are often available but you need to consider what parts of the documents to use, what their background is, and so on.
- Using information from the Internet requires consideration in regards to how best to turn it into data.
- In order to use existing data, you must consider the methodological steps and make decisions about what to sample, how to design the approach in a systematic way and how to analyze the data.

What's next

The chapter in the following book provides more detailed discussion of the issues covered in this chapter:

Corti, L. and Wathan, J. (2017) 'Online Access to Quantitative Data Resources', in N. Fielding, R. Lee and G. Blank (eds), *The SAGE Handbook of Online Research Methods*. London: Sage. pp. 489–507.

This next chapter addresses secondary analysis of data:

Radey, M. (2010) 'Secondary Data Analysis Studies', in B. Thyer (ed.) *The Handbook of Social Work Research Methods*. London: Sage. pp. 163–82.

This next source is a comprehensive overview of the field of social media research methods:

McCay-Peet, L. and Quan-Haase, A. (2017) 'What is Social Media and What Questions Can Social Media Research Help Us Answer?', in L. Sloan and A. Quan-Haase (eds), *The SAGE Handbook of Social Media Research Methods*. London: Sage. pp. 13–26.

This article addresses Big Data as a source for doing social research:

Tinati, R., Halford, S., Carr, L. and Pope, C. (2014) 'Big Data: Methodological Challenges and Approaches for Sociological Analysis', *Sociology*, 48(4): 663–81.

These two books give a good overview of the principles and pitfalls of analyzing documents. The first book includes a comprehensive account of analyzing them, as does the next one:

Tight, M. (2019) *Documentary Research in the Social Sciences*. London: Sage.
Rapley, T. (2018) *Doing Conversation, Discourse and Document Analysis*, 2nd edn. London: Sage.

This article addresses issues of analyzing documents in qualitative research:

Coffey, A. (2014) 'Analyzing Documents', in U. Flick (ed.) *The SAGE Handbook of Qualitative Data Analysis*. London: Sage. pp. 367–79.

The following case study can be found in the online resources. It should give you an idea of how a relevant problem is selected and turned into a research project using documents:

Chohan, U. W. (2019) 'Documentary Research: Positing Innovations in a National Budget Process', *SAGE Research Methods Cases*. doi: 10.4135/9781526469489.

This article, in the online resources too, describes the use of quantitative secondary data:

Garms-Homolová, V., Flick, U. and Röhnsch, G. (2010) 'Sleep Disorders and Activities in Long Term Care Facilities: A Vicious Cycle?', *Journal of Health Psychology*, 15 (5): 744–54.Using Existing Data

Research Methodology Navigator

Orientation

- Why social research?
- Worldviews in social research
- Ethical issues in social research
- From research idea to research question

Planning and design

- Reading and reviewing the literature
- Steps in the research process
- Designing social research

Method selection

- Deciding on your methods
- Triangulation and mixed methods

You are here in your project

Working with data

- Using existing data
- Collecting new data
- Analyzing data

Reflection and writing

- What is good research? Evaluating your research project
- Writing up research and using results

11

COLLECTING NEW DATA

You will:

- understand a range of social research methods for collecting data
- appreciate the similarities and differences, with regard to methods, between qualitative and quantitative research, and
- assess the methods available to you and what you can achieve with them.

In social research, there are three main forms of data collection – you can collect data by asking people (through surveys and interviews), observing, or studying documents. This chapter outlines each of these methods in turn.

Surveys and Interviews

By way of preparation for understanding the alternatives in speaking to people, please respond to the questionnaire given in the example in Box 11.1.

Box 11.1 Research in the real world

Questionnaire

Please fill in the following excerpt of a questionnaire for assessing the stress caused by studying, as per the following instructions:

'We ask you to answer some questions about your personal situation and about features of your university. Please circle without hesitation how far each of the areas referred to in the questions is stressful and satisfying for you. Very stressful is indicated by the value 5 in the stress scale, very satisfying by the value 5 in the satisfaction scale. Not stressful or satisfying is indicated by 1 on the respective scale. If one of the areas is only stressful or only satisfying for you, please circle only one of the scales (stressful or satisfying). If you circle 0, that means this area is neither stressful nor satisfying for you.'

Stressful										Satisfying
5	4	3	2	1	0	1	2	3	4	5

	Stressful									Satisfying	
Studying at this university limits me	5	4	3	2	1	0	1	2	3	4	5
Doing exams affects my studies	5	4	3	2	1	0	1	2	3	4	5
At university, social contact happens often	5	4	3	2	1	0	1	2	3	4	5
At university, social contact is rare	5	4	3	2	1	0	1	2	3	4	5
The professors and lecturers hardly bother about my studies or my work	5	4	3	2	1	0	1	2	3	4	5
Having leisure time affects my studies	5	4	3	2	1	0	1	2	3	4	5
Often, I have to decide between seeing family/ friends and studying	5	4	3	2	1	0	1	2	3	4	5
Sometimes I have to do things at university I do not agree with	5	4	3	2	1	0	1	2	3	4	5
Much is expected of me at university	5	4	3	2	1	0	1	2	3	4	5

Now – again by way of preparation – use the material in Box 11.2 to interview one of your (peer) students.

Box 11.2   Research in the real world

Interview about everyday life and studying

I would like to do an interview with you about the issue of 'everyday life and studying'. It is important that you answer according to your subjective point of view and express your own opinions. I will ask you several questions concerning situations in which you have made experiences of studying and ask you to recount those situations for me.

First of all, I would like to know:

1   What was your first experience of studying here? Could you give me a concrete example of this?
2   Please tell me about the course of your day yesterday and when your studies played a role in it.
3   When you look at what you do during your time off, what role does your study play in it? Please give me a concrete example of this.
4   When looking at your life in general, do you have the feeling that studying takes up a bigger part of it than you expected? Can you outline this by giving me an example?
5   What do you link to the word 'stress'?
6   When you look back, what was your first experience of stress as a student? Could you tell me about this situation?

What do you find are the differences between these two experiences? Which exercise do you feel would cover the respondent's situation better? Which form gave more room for the respondent's own views? Which one do you think produces the clearer data?

These two research instruments, in Boxes 11.1 and 11.2, are adapted from a questionnaire often used to study stress in the workplace and from an episodic interview (see below). They represent contrasting forms of asking people about their situation. According to the worldviews behind them, they are more interested in testing hypotheses (e.g. in the questionnaire that a lack of social contacts is a stressor) or in exploring the individuals' subjective experiences (e.g. in the interviews, the first memories of university). The main difference lies in the degree of standardization of the procedure – the questionnaire comes with a predefined list of questions and answers, while the interview is more open-ended. In the interview, the questions can be varied in their sequence and interviewees can use their own words and decide what they want to mention in their responses. Below, we discuss both methods in more detail.

Quantitative Surveys: Questionnaires

Most surveys are based on questionnaires. These may be answered either in written form or orally in a face-to-face interrogation with a researcher noting down the answers. A characteristic of questionnaires is their extensive standardization. The researchers will determine the formulation and sequencing of questions and possible answers. Sometimes a number of open or free text questions, which the respondents may answer in their own words, are included too.

Questionnaire studies aim at receiving comparable answers from all participants. Therefore, the questions as well as the data collection situation are designed in an identical way for all participants. When one is constructing a questionnaire, rules for formulating questions and arranging their sequence should be applied.

Question wording

Here, there are three main issues: how to formulate questions, which kinds of question and possible answers are appropriate, and the purpose of asking the questions. Questions should collect, directly or indirectly, the respondents' reasons for a specific behavior or attitude and show their level of knowledge concerning the issue under examination.

How instructive the information received is will depend both on the type of question and on its position in the questionnaire. It will also depend on whether or not (a) the questions fit the respondent's frame of reference and (b) the respondent's and the interviewer's frames of reference correspond. Finally, the situation, in which a question is asked and should be answered, plays a major role.

If you wish to identify the interviewee's frame of reference, you should ask for reasons to be given (e.g. 'Why did you choose this topic for your studies?'). You will need to take the interviewee's level of knowledge into account: if questions are too complex, they are likely to be misunderstood and produce diffuse answers. Accordingly, you should translate complex issues into concrete, clearly understandable questions. Where possible, use colloquial language. Note here that it may not make much sense to incorporate the language of your hypothesis directly into your questions.

Be careful to avoid multidimensional questions, for example: 'How and when did you discover that ...?'. If questions include several dimensions, their comparability will be reduced, because respondents may pick up on different dimensions in the questions. Also, avoid questions with a bias, which suggest a specific answer (suggestive questions), or with certain assumptions. For example, if you are studying burnout processes, the question 'How burned out by your job do you feel?' assumes that the respondent already feels burned out. Instead, you should find out through a prior question whether this is in fact the case. Questions should be as short and simple as possible and aligned to the respondent's frame of reference as closely as possible. Double negations should be avoided. So too should unclear or technical terms. For example, rather than ask 'What is your subjective definition of health?', ask 'What do you associate with the word "health"?'. In formulating your questions, you should try to get close to your participants' vocabulary range and normal parlance and to select words used in the question carefully to avoid bias. For example, you should ask 'how challenging is your job?' rather than 'how stressful is your job?' The more respondents are familiar with the wording of a questionnaire, the more likely they will identify with the questions and relate them to their personal situations or experiences.

You should also recognize that the features of the event you are studying will influence the quality of the answers provided. The further in the past an event is, the less exact the answers will be. The more respondents are interested in an issue, the more detailed and accurate the answers will be. The more frightening an event has been for respondents, the more likely it is they will have forgotten it. The more something is linked to social rejection (e.g. time spent in psychiatric wards), the less a person will talk or make statements about it. The higher the value of something (e.g. income, status), the more likely it is that responses will include overestimates. For closed questions with two possible answers, it is more likely that the second will be given. For questions with a list of possible answers ('list questions'), a person who does not know the answer is likely to select an answer from the lowest third of the list: this is known as the 'position effect'. These and other issues around how to formulate and position a question are dealt with in detail by Neuman (2014, Chapter 10).

Some texts (e.g. Bortz and Döring 2006, pp. 244–6) provide checklists for assessing standardized oral or written questions. Such checklists include guiding points such as:

- Are all the questions necessary?
- Does the questionnaire include redundant questions?
- Which questions are superfluous?
- Are all the questions formulated easily and clearly?
- Are there negative questions, with answers that could be ambiguous?

- Are the questions formulated too general?
- Will the interviewee be potentially able to answer the questions?
- Is there a risk that the questions will be embarrassing for the interviewee?
- Might the result be influenced by the position of the questions?
- Are questions formulated in a suggestive way?
- Are the opening questions adequately formulated and has the questionnaire a properly reflected end?

Types of questions and possible answers

Questions can be distinguished according to how they can be answered. Open questions in a questionnaire do not come with answers defined in advance, while closed questions already specify the alternatives for answering them. Such alternatives may be specified in the wording of the question (e.g. 'Are most nurses satisfied or unsatisfied by their work?'), which permits only a limited number of answers (in our example, two). A closed question may limit the number of possible answers by use of a scale of agreement, as shown in Figure 11.3.

An alternative is to present different possibilities for answering, as illustrated in Box 11.3.

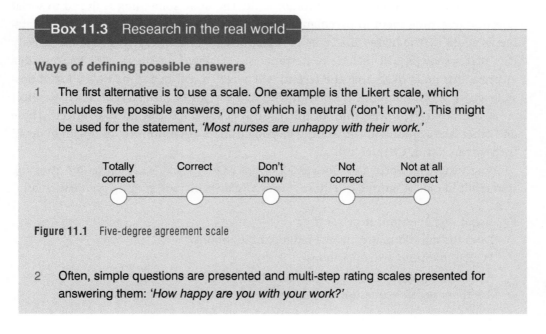

Box 11.3 Research in the real world

Ways of defining possible answers

1   The first alternative is to use a scale. One example is the Likert scale, which includes five possible answers, one of which is neutral ('don't know'). This might be used for the statement, *'Most nurses are unhappy with their work.'*

Totally correct	Correct	Don't know	Not correct	Not at all correct
◯	◯	◯	◯	◯

Figure 11.1   Five-degree agreement scale

2   Often, simple questions are presented and multi-step rating scales presented for answering them: *'How happy are you with your work?'*

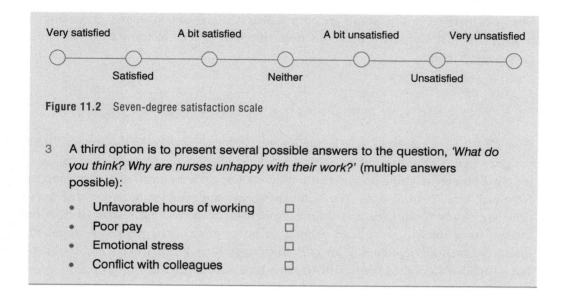

Figure 11.2 Seven-degree satisfaction scale

3   A third option is to present several possible answers to the question, *'What do you think? Why are nurses unhappy with their work?'* (multiple answers possible):

- Unfavorable hours of working ☐
- Poor pay ☐
- Emotional stress ☐
- Conflict with colleagues ☐

It is also possible to present a situation. Here you would list several possible answers to the question, 'What would you do in such a situation?' Or you could ask questions relating to how some other person would react in the situation. Sometimes questionnaires include control questions, which re-address issues in earlier questions using different wording.

		Degree of agreement				
		Not at all				Full
A	Brexit will be able to solve the economic problems of the United Kingdom	1	2	3	4	5
B	Brexit will lead to a separation of Scotland from the United Kingdom	1	2	3	4	5
C	Brexit will do more harm to the United Kingdom than to the European Union	1	2	3	4	5

Figure 11.3 Five-step answer scale

Positioning of questions

Questions may influence each other. That is, how one question is answered may be influenced by the question asked immediately before it. This is known as the 'halo effect'.

The links between two questions are not necessarily designed deliberately. In general, it is suggested that you group questions from the most general at the start to the most concrete at the end.

Suggestions for how to formulate questions

Porst (2014) has formulated '10 commandments' of question wording for surveys. These are given in Box 11.4. His starting point is as follows: 'Bad questions lead to bad data and no procedure of weighing and no method of analyzing the data can make good results from bad data' (2014, p. 118). The 'commandments' are designed to ensure that questions are clearly formulated and do not over-challenge or confuse respondents. Porst intends these rules to provide an orientation only: 'most of the rules leave room for interpretation and sometimes ... are even in competition with each other and thus cannot be applied a hundred percent at the same time' (2014, p. 118).

Box 11.4 Research in the real world

Suggestions for how to formulate questions

Porst's (2014) '10 commandments' of question wording for surveys include the following points: questions should be based on 'simple unambiguous concepts which are understood by all respondents in the same way'; ... one should 'avoid long and complex, ... (or) hypothetical questions, ... double stimuli and negotiations, ... presumptions and suggestive questions or aiming at information many respondents presumably will not have'; one should 'use questions with a clear temporal reference, ... answer categories which are exhaustive and ... free of overlap ... ensure that the context of a question does not impact on answering it ... [and] ... define unclear concepts' (2014, pp. 99–100).

Porst also gives examples of how to use these commandments to distinguish bad from good questions. For his first commandment ('You shall use simple unambiguous concepts which are understood by all respondents in the same way!'), a badly formulated question would be: 'What do you think, will the economic cycle in Germany at the end of 2013 compared to today be very positive, rather positive, rather negative or very negative or will all remain as it is today?' Better wording for this question would be: 'What do you think: How will the economic situation be in Germany at the end of 2013: much better than today, a little better, equal or much worse?' (2014, p. 100). For the commandment 'You shall avoid long and complex questions!' Porst sees the following as an example of poor wording: 'As you know, some people are politically rather active, while other people do not find the time, or have no interest in actively engaging in politics. I will now read a number of things that people do. Please tell me, for each issue, how often you personally do the thing or how often it occurs for you (list with response categories: often, sometimes, rarely, never). First, how often do you engage in a political discussion?' (2014, p. 103). Porst suggests that better wording for this would be: 'How often do you take part in public discussions about political issues: often, sometimes, seldom or never?' (2014, p. 104).

The 'commandments' give some guidance for wording your survey questions in a more easily accessible way. The examples illustrate this direction, although the suggestions Porst provides for better wording are not necessarily the optimal solution; rather, they flesh out a way of nailing down your interests into simple and easily understandable questions.

Practical issues

Often, questionnaires are distributed and recipients are asked to send them back within a certain time frame. In this case, the questionnaire will need to be accompanied by a letter providing sufficient information about the study, explaining its importance and encouraging the recipient to participate. A major issue is the response rate (i.e. the number of questionnaires which are actually filled in and sent back). It is not only necessary that enough questionnaires are returned (50% would be quite a good ratio here), but, in addition, that enough of those returned have been filled in *completely*.

Finally, you should check how far the distribution of the returned questionnaires corresponds to the distribution in the original sample. To use a simple example: the proportion of men to women in a population is two to one. The selection of the sample and the mail-out of the questionnaires have taken this proportion into account. If the proportion that characterizes the distribution in the returned questionnaires is completely different from the proportion in the original sample, the results of the study can be generalized to the population only in a very limited way. An alternative is to decide not to mail out the questionnaires, but instead to do a standardized survey with an interviewer who visits the participants. This may improve the return of questionnaires, though it will be much more time-consuming and expensive.

Example: A youth health survey

The youth health survey has already been described in Chapter 4. It is based on a questionnaire covering the topics in Box 11.5 (see Richter 2003).

> **Box 11.5** Research in the real world
>
> **Topics in the youth health survey**
>
> - Demographic information: gender, age, family structure, location, form and grade of school, socioeconomic status, ethnic background.
> - Subjective health: psychosomatic complaints, mental health, allergies, life satisfaction, etc.
> - Risk of accidents and violence: accidental injuries, mugging (perpetrator and victim), participation in fights, use of substances (tobacco, alcohol, illegal drugs).
> - Eating behavior and diet: eating habits, body-mass-index diet, etc.
> - Physical activity: sport, physical effort, inactivity due to television and computer use, etc.
> - Social resources: number of friends, support of parents and peers, family situation, living conditions, etc.
> - School: performance requirements, quality of teaching, support of parents, fellow students and teachers, friends at school, involvement in school life, etc.
> - Peer group and leisure activities: frequency of meetings, use of media, membership of associations and organizations, etc.

Summary

Questionnaires are appropriate for a study when (a) you want to test a hypothesis with your empirical work, (b) knowledge of the issue allows you to formulate a sufficient number of questions in an unambiguous way, and (c) a large number of participants will be involved. Questionnaires are highly standardized. The sequence and formulation of questions are defined in advance, as are the possible answers. Questionnaires are often sent out which affects response rates.

Figure 11.4 illustrates the varying degrees of openness for participants to use their own formulations and words within several methods of interviewing.

Qualitative Inquiries: Interviews and Focus Groups

We turn now to consider various forms of interviews. There are semi-structured interviews: these are based on an interview guide with (sometimes different types of) questions to be answered more or less openly and extensively. There are expert interviews. There

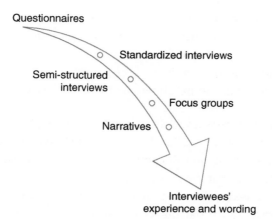

Figure 11.4 Degrees of interviewees' freedom in using their own words and formulations

are also narrative-based interviews, which focus on inviting interviewees to recount (aspects of) their life history. We also find interviews mixing both questions and narrative stimuli. Finally, there are interviews involving groups rather than single participants. Focus groups are another form of collecting verbal data. Below we consider some of these types in more detail.

Semi-structured interviews

For semi-structured interviews, a number of questions are prepared that between them cover the intended scope of the interview. For this purpose, you will need to develop an interview guide as an orientation for the interviewers. In contrast to questionnaires, interviewers can deviate from the sequence of questions. They also do not necessarily stick to the exact formulation of the questions when asking them. The aim of the interview is to obtain the individual views of the interviewees on an issue. Thus, questions should initiate a dialogue between interviewer and interviewee. Again, in contrast to questionnaires, in an interview you will not present a list of possible answers. Rather, the interviewees are expected to reply as freely and as extensively as they wish. If their answers are not rich enough, the interviewer should probe further.

In constructing an interview guide and conducting the interview, four criteria are helpful. Merton and Kendall (1946) provided these criteria initially for focused interviews. They emphasise:

- non-direction in relations with the interviewee
- specificity of the views and definition of the situation from the interviewees' point of view
- covering a broad range of meanings of the issue
- the depth and personal context shown by the interviewee.

For this purpose, in most cases a variety of questions are applied. Questions may be open ('What do you link to the word "health"?') or semi-structured ('If you think about what you eat, what role does health play in this context?'). Only rarely are structured questions used, which present a statement that the interviewee is expected to agree with or to reject (e.g. 'In your everyday life as a teacher, do you have the impression that many children are eating in too unbalanced a way?'). Such statements almost have a suggestive character, though they are used sometimes in this context to stimulate interviewees to reflect on their position and maybe also to make them explicitly express their distinction from the question or statement. Other elements in a semi-structured interview – such as presenting a fictitious case story with questions referring to it (also known as the 'vignette' technique) – can have a similar function.

Decisive in the success of a semi-structured interview is that the interviewer probes at apposite moments and leads the discussion of the issue into greater depth. At the same time, interviewers should ask all the questions in the interview that are relevant to the issue. Open questions should allow room for the specific, personal views of the interviewees and also avoid influencing them. Such open questions should be combined with more focused questions, which are intended to lead the interviewees beyond general, superficial answers and also to introduce issues that the interviewees would not have mentioned spontaneously. The construction of an interview should of course be linked closely to the aims and target group of the research.

Expert interviews

Expert interviews (see Bogner et al. 2009) focus less on the interviewees' personalities and more on their expertise in a specific area. In contrast, if you interview patients about their illness, the people themselves and their personal experiences will be of greater interest. Accordingly, in the first case the experts will be asked more focused questions, while in the second case the patients will be asked more open questions. Expert interviews can be used with different aims. Bogner and Menz (2009, pp. 46–9) suggest three types of expert interviews. Such an interview can be used for exploration: (1) for orientation in a new field in order to give the field of study a thematic structure and to generate hypotheses (2009, p. 46). This form of interview can also be used to prepare the main instrument in a study for other target groups (e.g. patients). The systematizing expert interview (2) can be applied to collect context information complementing insights coming from applying other methods (e.g. interviews with patients). Theory-generating expert interviews (3) aim at developing a typology or a theory about an issue from reconstructing the knowledge of various experts – for example, about contents and gaps in the knowledge of people working in certain institutions concerning the needs of a specific target group. In this context, the distinction made by Meuser and Nagel (2002, p. 76) becomes relevant: both process knowledge and context knowledge can be reconstructed in expert interviews. In the former, the aim is to have information about a specific process: how does the introduction of a quality management instrument in a hospital proceed; which problems occurred in concrete examples; how were they addressed? What happens if people in a specific situation (e.g. being homeless) become chronically ill: whom do they address first; which barriers do they meet; how does the typical patient career develop? From such a process knowledge, we can distinguish context knowledge: how many of these cases can be noted; which institutions are responsible for helping them; which role is played by health insurance or by the lack of insurance, etc.?

Furthermore, Bogner and Menz (2009, p. 52) distinguish various forms of knowledge that can be addressed in expert interviews: (1) *Technical knowledge* 'contains information about operations and events governed by rules, application routines that are specific for a field, bureaucratic competences and so on'. (2) *Process knowledge* 'relates to inspection of and acquisition of information about sequences of actions and interaction routines, organizational constellations, and past or current events, and where the expert because of his or her practical activity, is directly involved or about which he or she at least has more precise knowledge because these things are close to his or her field of action'. (3) *Interpretive knowledge* includes 'the expert's subjective orientations, rules, points of view and interpretations'. This can be accessed in expert interviews but should be seen 'as a heterogeneous conglomeration' (p. 52). The interview guide in expert interviews has much more a directive function with regard to excluding unproductive topics given the narrow focus in its application and due to time pressure in the interview. On this point, Meuser and Nagel discuss a series of problems and sources of failure in expert interviews. The main question is whether or not the interviewer manages to restrict and determine the interview and the interviewee to the expertise of interest. Meuser and Nagel (2009, pp. 71–2) identify a number of ways that an expert interview can fail:

- The expert blocks the interview in its course, because he proves not to be an expert for this topic as previously assumed.
- The expert tries to involve the interviewer in ongoing conflicts in the field and talks about internal matters and intrigues in his work instead of talking about the topic of the interview.

- The expert often changes between the roles of expert and private person, so that more information results about her as a person than about her expert knowledge.
- As an intermediate form between success and failure, the 'rhetoric interview' is mentioned. In this, the expert gives a lecture on his knowledge instead of joining the question–answer game of the interview. If the lecture hits the topic of the interview, the latter may nevertheless be useful. If the expert misses the topic, this form of interaction makes it more difficult to return to the actual topic of relevance.

Interview guides have two functions:

> The work, which goes into developing an interview guide, ensures that researchers do not present themselves as incompetent interlocutors … The orientation to an interview guide also ensures that the interview does not get lost in topics that are of no relevance and permits the expert to extemporize his or her issue and view on matters. (Meuser and Nagel, 2002, p. 77)

The expert interview can be used as a stand-alone method if your study aims at comparing the contents and differences of expert knowledge held by representatives of different institutions in a field. In this case, you will select enough relevant persons in a sufficient variety to interview them. But in many studies, expert interviews are used to complement other methods – beforehand to develop the main instrument or for an orientation in the field (see above), or in parallel to round up information from other interviews. Expert interviews can also be used as a complementary method after the main data collection, such as in an expert validation of findings resulting from other interviews. Both can be seen as an example of triangulation (see Chapter 9) of different perspectives on an issue under study.

In all applications of semi-structured interviews, it is only in the interview situation that you can decide when and how extensively to probe. For this purpose, it can be helpful first to do interview training before data collection. This will consist of role plays of practice interviews and of comments by others.

Narrative interviews

A different path to discovering the subjective views of participants is found in the narrative interview. Here, it is not questions that are central. Instead, interviewees are invited to present longer, coherent accounts (say, of their lives as a whole or of their disease and its course) in the form of a narrative. The method is prominent in biographical research. Hermanns describes its basic principle of collecting data as follows:

> In the narrative interview, the informant is asked to present the history of an area
> of interest, in which the interviewee participated, in an extempore narrative ... The
> interviewer's task is to make the informant tell the story of the area of interest in
> question as a consistent story of all relevant events from its beginning to its end.
> (1995, p. 183)

The narrative interview comprises several parts – in particular (a) the interviewee's main
narrative, following a 'generative narrative question', (b) the stage of narrative probing
in which narrative fragments, that were not exhaustively detailed before, are completed,
and (c) the final stage of the interview (known as the 'balancing phase'), consisting of
questions that take the interviewees as experts and theoreticians of themselves.

If you aim to elicit a narrative that is relevant to your research question, you must
formulate the generative narrative question broadly, yet at the same time sufficiently
specifically, to produce the desired focus. The interest may relate to the informant's life
history in general. In this case, the generative narrative question will be rather generalized –
for example, 'I would like to ask you to begin with your life history.' Or the interest may
lie in some specific, temporal and topical aspect of the informant's biography, such as a
phase of professional reorientation and its consequences. An example of the type of generative question required here is:

> I want to ask you to tell me the story of your life. The best way to do this would
> be for you to start from your birth, with the small child that you once were, and
> then tell me all the things that happened one after the other up to today. You can
> take your time in doing this, and also give details, because for me everything is of
> interest that is important for you. (Hermanns 1995, p. 182)

It is important to ensure that the first question really is a narrative generative question
and that the interviewer does not impede the interviewee's storytelling with questions
or directive or evaluating interventions. A major test of validity here is the question of
whether a narrative really was presented by the interviewee. Here, you should take into
account that 'It is always only "the story of" that can be narrated, not a state or an always
recurring routine' (1995, p. 183).

There is a problem presented here by the role expectations of the parties involved.
They include a systematic violation of the usual expectations surrounding an 'interview',
because the main part of the narrative interview does not consist of asking questions in
the traditional way. At the same time, expectations pertaining to the 'everyday narrative'
are also violated, as the extensive space for narration that the interviewee is given here
in a unilateral way hardly ever occurs in everyday life. This can produce difficulties for
both parties involved. As for other forms of interviewing, interview training focused on

active listening and a clarification of the specific character of the interview situation for the interviewee is necessary.

The narrative interview aims to access interviewees' subjective experiences through three means, namely:

1   The opening question, which aims to stimulate not simply a narrative, but specifically a narrative about the topical area and period of interest in the interviewee's biography.
2   The orientation on the scope for the interviewees in the main narrative part, which enables them to tell their stories perhaps for several hours.
3   The postponing of concrete, structuring, thematically deepening interventions in the interview until the final part, when the interviewer is supposed to pick up issues briefly mentioned and to ask focused questions. This means that any structuring activities of the interviewer are located at the end of the interview and even more so at the beginning.

The episodic interview

The interviews presented so far choose one approach – either questions and answers, or narrative stimulus and telling life histories – as the major approach to an issue under study. For many research questions, however, it is necessary to use a method that combines both principles – narrative and interrogation – equally. The episodic interview (see Flick 2018a; 2018b) starts from the assumption that individuals' experiences about a certain area or issue are stored in the forms of narrative-episodic and semantic knowledge. While the first form is focused closely on experiences and linked to concrete situations and circumstances, the second form of knowledge contains abstracted, generalized assumptions and connections. In the first case, the course of the situation in its context is the central unit, around which knowledge is organized. In the second case, concepts and their interrelations form the central units. In order to cover both parts of knowledge about an issue, the method collects narrative-episodic knowledge in narratives, and semantic knowledge in concrete and focused questions. Figure 11.5 illustrates the forms of knowledge addressed in the episodic interview.

The aim is systematic connection of the two kinds of data (i.e. narratives and answers) and thus of the two kinds of knowledge they make accessible. The episodic interview gives space for context-related presentations in the form of narratives, as they address experiences in their original context more immediately compared to other forms of presentation. At the same time, narratives can elucidate more about processes of constructing

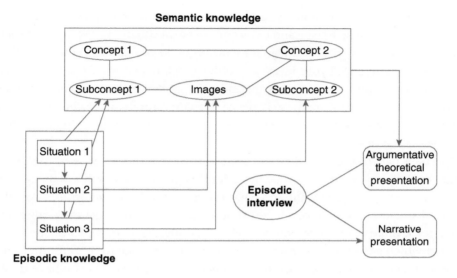

Figure 11.5 Forms of knowledge in the episodic interview

realities on the part of interviewees than other approaches that focus on more abstract concepts and answers.

The focus in the interview is on situations and episodes in which the interviewee has had experiences that are relevant to the research issue. Which form of presentation (description or narrative) is chosen for the particular situation, as well as the selection of situations, can be decided by the interviewees according to their subjective relevance to the issue. The aim of the episodic interview is to permit the interviewee for each substantial area to present experiences in a general or comparative form, and at the same time to recount relevant situations and episodes. Planning, doing and analyzing episodic interviews involve the following steps (see Flick 2000a). As a preparation, you should develop an interview guide, which includes narrative stimuli and questions about the areas and aspects you see as relevant to the research issue. At the beginning of the interview, you should introduce the principle of the interview, explaining that you will repeatedly invite the interviewee to recount specific situations. The general concern around interviews based on narratives is relevant here, too; some people have more problems with narrating than others. Therefore, it is very important that you explain the principle of recounting situations to the interviewees. Interviewees should be encouraged to recount relevant situations, rather than merely mention them. You should take care that the interviewee has understood and accepted this method. An example of such an introduction to the interview principle is: 'In this interview, I will ask you a number of times to tell me about situations in which you have had experience with the topic of health.'

With the opening questions, you invite the interviewees to present their subjective definition of the research issue and to recount relevant situations. Examples of such questions are: 'What does "health" mean for you? What do you link with the word "health"?' or 'What has influenced your ideas of health in a particular way? Could you please tell me about a situation which makes this clear for me?' Which exemplary situation they choose to recount for this purpose is always a decision for the interviewees. You can subsequently analyze both the selection and the account of the situation that each interviewee focuses on.

The next part of the interview will focus on the role or relevance of the issue under study for the participants' everyday lives. For this purpose, you can ask them to recount how a day (e.g. yesterday) went and what relevance the issue had on that day. If a multitude of situations is mentioned, you should pick up on the most interesting ones and probe for a more detailed response. Again, it is interesting in terms of the analysis to identity which situations the interviewees themselves choose to recount. An example of a question addressing changes in the relevance of the issue is: 'Do you have the impression that your idea of health has changed during your professional life? Could you please recount a situation for me which demonstrates this change?'

Next, you will ask the interviewees to outline their personal connection to major aspects of the study topic. For example, 'What does it mean for you to promote health as part of your work? Could you please tell me about a situation of this kind?' You should ask the interviewee to illustrate these examples of personal experiences and subjective concepts as substantially and comprehensively as possible by probing where necessary.

In the final part of the interview, you will ask the interviewees to talk about more general aspects of the focal issue and to present their personal views in this context. This is meant to extend the scope of the interview. You should try as much as possible to link these general answers with the more personal and concrete examples given before so that possible discrepancies and contradictions become visible. An example of such a generalizing question is: 'Who do you think is responsible for your health? Is there a situation you can think of in which you can outline this?'

As in other interviews, the final question should ask the respondent about aspects missing from the questions or whether there is anything that should be added (e.g. 'Is there anything else ...?').

Immediately after the interview, you should complete a sheet on which you document socio-demographic information about the interviewee and features of the particular interview situation. Any disturbances and anything mentioned about the subject of the interview after the recording device was turned off should be noted as well, provided that is ethically acceptable (see Flick 2018a, p. 437 for an example of such a documentary sheet).

Example: Health concepts of homeless adolescents

In Chapter 4, our study on the health concepts of homeless adolescents was described. Our interviews with the adolescents were based on the episodic interview and covered the areas in Box 11.6, for which some examples of questions and narrative stimuli are given.

Box 11.6 Research in the real world

Excerpts from an interview guide for an episodic interview

Health concepts: general and individual definition of health, relevance and possible influences

- First of all, I would like to ask you to tell me what your family life was like. What made you turn to living on the street? Can you tell me about a situation which explains this for me?
- Now let's turn to our topic. What is that for you – health?
- How do you decide that you are healthy? Can you tell me about a situation which helps me to understand this?
- Links of factors and risks in the life world to health concepts and practices
- If you think about how you live these days, is there anything which influences your health? Can you tell me about a situation which explains this for me?
- Do you see any relation between your current housing situation and your health? Can you tell me about a situation which explains this for me?
- How do you deal with illness? Can you tell me about a situation which explains this for me?
- Do you think that your financial situation has an influence on your health? How? Can you give me an example of this?

Risk behavior and secondary prevention as part of health practices

- Do you think that you sometimes risk your health by what you do? Is there a situation here you can tell me about?
- In terms of drugs and alcohol, what do you take? Please be specific.

Experiences of illness and dealing with them

- When you do not feel well, what are your major concerns? How do you think this problem [mentioned by the interviewee] comes about?

Focus groups and group discussions

An alternative to interviewing individuals is to conduct group interviews in which several participants are asked the same question and they answer one after the other. Alternatively – and more commonly – a group may be used as an interactive community. Since the 1950s, group discussions have been used mainly in German-speaking areas, with focus groups used, in parallel, in the Anglo-Saxon world. In both cases, data collection is based on initiating a discussion in a group about the research issue. For example, pupils discuss their experiences of violence and how they deal with it, or students evaluate the quality of their courses and of the teaching in focus groups.

The starting point for using this method is that these discussions can make apparent how attitudes or evaluations have developed and changed. Participants are likely to express more and go further in their statements than in single interviews. The dynamics of the groups become an essential part of the data and of their collection.

For group discussions, a variety of forms of group can be used. You can start from a natural group (i.e. a group that exists in everyday life beyond the research), from a real group (which is confronted in its everyday life by the research issue) or from an artificial group (set up for the research according to specific criteria). You can also distinguish between homogeneous and heterogeneous groups. In homogeneous groups, the members are similar in essential dimensions with respect to the research question: for example, they may have a similar professional background. In heterogeneous groups, the members should differ in the features that are relevant for the research question. Focus groups and group discussions can be moderated in different ways. You can refrain from any form of moderation or do some formal steering (setting up a list of speakers) or a more substantial moderation (by introducing topics or intervening with some provocative questions). One starts a focus group or a group discussion with a discussion stimulus, which can be a provocative question, a comic or a text, or the presentation of a short film (see also Barbour 2018; Morgan and Hoffman 2018).

In general, you should be aware that running focus groups may require a lot of organizational effort (in coordinating the date of the meeting, for example). Afterwards, it is the various groups that can be compared rather than all participants across the groups. Thus, you should compare the various groups with each other, rather than the single members. You should use this method only if you have good reason to because of the research question and not because you expect to save some time compared to single interviews (see also Chapter 13).

Online Surveys, Interviews and Focus Groups

The three basic methods for gathering participants' statements – namely drawing on opinions, stories and other forms of verbal data – have been transferred to online research.

Online surveys

There are several assessments of online surveys. Most research using the Internet is quantitative and consists of online surveys, web-based questionnaires or Internet experiments (see Hewson et al. 2003). It has been estimated that at least a third of all surveys worldwide are online surveys (Evans and Mathur 2005, p. 196) and the trend is upwards. Bryman (2016) discusses e-mail surveys and web surveys. The first, as the name suggests, is sent by e-mail to recipients selected in advance. The questionnaire is attached to an e-mail with an expectation that the recipients answer the questions and send the questionnaire back as an attachment to their reply e-mail. An alternative is to send the questionnaire embedded in the e-mail itself, with the answers given by adding an 'x' to the questions or by writing text into the e-mail and sending it back by touching the reply button. Although the latter may be easier to handle for the respondent, some e-mail programs may have problems with the formatting. Web surveys are more flexible in formatting the whole questionnaire and the answer options. Questionnaires can be designed appealingly and it is easier to include filter questions or skip questions (the answer directs the participant to different questions to refer to next, or questions are left out after a specific answer). Participants may be addressed by placing a banner similar to an advertisement on a webpage and respondents can click on the questionnaire and fill it in. There are also software tools and services available on the Internet which can facilitate doing your online survey; many companies offer these services professionally (which means you have to pay for them). If you search 'Google survey', you will end up with a long list of services such as Survey Monkey and Bristol Online Surveys (www.onlinesurveys.ac.uk).

Compared to postal questionnaire surveys, online surveys have a number of advantages. These include:

- *Low cost* – as you will not have to print your questionnaire, you can save money on printing, envelopes and stamps. The questionnaires you receive back are already 'in' the computer and more easily transferred into the statistics software.
- *Time* – online questionnaires come back more quickly than postal questionnaires.

- *Ease of use* – online questionnaires are easier to format and easier to navigate for the participant (see above).
- *Lack of spatial restrictions* – you can reach people over long distances without waiting for the questionnaire to travel back and forth.
- *Response rate* – the number of unanswered questions in most cases is lower with online surveys, while open questions tend to be answered in a more detailed way and the answers are already given in a digital format.

However, there are also a number of disadvantages of online questionnaires compared to traditional surveys. The response rates can be lower in some cases; you will only reach populations that are already online. There is skepticism about anonymity, especially on the part of potential participants; this may reduce the motivation to respond. Alternatively, people sometimes reply more than once (see Hewson et al. 2003, p. 43; Bryman 2016, p. 192).

Box 11.7   Research in the real world

Longitudinal monitoring of students' quality assessments

At the University of Potsdam, an online panel was established to monitor student biographies, course evaluations and assessments of the study program quality, in a longitudinal perspective. Students were invited to join the online panel through online and offline contacts. Students were contacted before they started their studies and before enrolling for the programs, and were motivated by a lottery for participants. Nevertheless, of the 18,000 students at the university, only 700 were still in the panel after a year. The research team took careful precautions to guarantee the anonymity of the participants and to be able to link data from the several waves of the survey (see Pohlenz et al. 2009 for details).

Sue (2007, pp. 20–1) has outlined a research timeline for online surveys to be completed in 13 weeks. The first three weeks are devoted to writing the study objectives, reviewing the literature and revising the objectives. The next three weeks are devoted to selecting survey software and developing, pre-testing and revising the questionnaire. In week 7, the actual data collection is run and incoming responses are monitored. Weeks 8 and 9 focus on reminding non-respondents and analyzing the data. In the remaining four weeks, you will write first and second drafts of research reports and present them. This timeline looks very tight, in particular in the part where the research is run – but it at least gives an initial indication of how to progress.

Online interviews

You can organize online interviewing in several ways. You may use a *synchronous* form. This means that you contact your participant while you are both online at the same time – such as in a chat room where you can directly exchange questions and answers. This comes closest to the verbal exchange in a face-to-face interview. Alternatively, you can organize online interviews in an *asynchronous* form, which means you send your questions to the participants and they send their answers back later; here you are not necessarily online at the same time. The latter version is mostly done through e-mail exchanges and comes close to what you do in a questionnaire study. You can also use messaging services like Skype to establish an *immediate* dialogue in the format of question and answer. If the technical resources are available, you can even establish a video dialogue in which you see your respondent and vice versa.

Each of these alternatives needs technical resources (like a camera, a fast broadband connection and the like) on both sides. Mann and Stewart (2000, p. 129), following Baym (1995), see five questions as important to consider for computer-mediated interaction in interviews. They are:

1   What is the purpose of the interaction/interview? This will influence the interest of possible participants in whether or not to become involved in the study.
2   What is the temporal structure of the research? Are synchronous or asynchronous methods used, and will there be a series of interactions in the research or not?
3   What are the possibilities and limitations of the software which will influence the interaction?
4   What are the characteristics of the interviewer and the participants? What about their experience of and attitude to using technology? What about their knowledge of the topics, writing skills, insights, etc.? Is one-to-one interaction or researcher– group interaction planned? Has there been any interaction between researcher and participant before? How is the structure of the group addressed by the research (by hierarchy, gender, age, ethnicity, social status, etc.)?
5   What is the external context of the research – the inter/national culture and/ or communities of meaning that are involved? How do communicative practices outside of the research influence the research itself?

If you conduct your interviews in an asynchronous form, the delay between question and answer may influence the quality of your data and the thread of the interview may get lost. When running the interview itself, you can either send one question or a couple of questions, wait for the answers and then probe (as in a face-to-face interview), or continue sending the next questions. If there is a long delay before answers come, you can send a reminder (after a few days, for example). Bampton and Cowton (2002) view a decline in the length and quality of responses, as well as a tendency for answers to come more slowly, as a sign of fading interest on the part of the participant and a signal for the interview to come to an end.

The advantages mentioned for online interviewing are the same as for online research in general. You may save time and costs and can reach people over great distances. An additional advantage is the greater anonymity for the participant, in particular in e-mail interviews. If you choose a video call, the anonymity for the participant is no longer given as it is in the e-mail interview, where statements are simply exchanged as text. At the same time, you can more easily contextualize statements in paralinguistic contexts like facial expressions. The disadvantages are doubts about 'real' identities (who am I talking to), the loss in direct rapport with the participants and problems in probing when answers remain unclear. The latter applies more to e-mail interviews, which come closer to the situation of filling in a questionnaire (see Salmons 2010 for further detail on planning and doing online interviews in a synchronous way).

Online focus groups

There has been particular interest in online focus groups. Again, one can distinguish between synchronous (or real-time) and asynchronous (non-real-time) groups. The first type of online focus group requires that all participants are online at the same time. They may participate via a chat room or by using specific conferencing software. The latter option requires that all participants have this software on their computers or be provided with it. Besides the technical problems this may cause, many people may hesitate to receive and install software for the purpose of taking part in a study. Asynchronous focus groups do not require that all participants are online at the same time (and this prevents problems of coordination).

For online focus groups to work, ready access for participants is required. Mann and Stewart (2000, pp. 103–5) describe in some detail the software you can use to set up a synchronous focus group ('conferencing software'). They also describe how to design websites, and how these can facilitate access for intended participants and exclude those not intended to have access. The authors also discuss how the concepts of naturalness and neutrality concerning the venue of a focus group also apply online. For example, it is important that participants can take part in the discussions from their computers at home or at their workplace, rather than from a special research site. As a start, it is important to create a welcome message, which invites the participants, explains the procedures and what is expected of participants, and describes the rules of communication among participants (e.g. 'please be polite to everyone'), and so on (see Mann and Stewart 2000, p. 108 for an example). The researcher should – as with any focus group – create a permissive environment.

The advantages of doing focus groups online are that you may save time and costs, as you will not have to set up your group at the same time and place and you will save transcription costs as data already come in digital form. Other advantages are that there is greater anonymity for participants and that group dynamics play a smaller role compared to real-world focus groups: it is less likely here that participants will dominate the group or suppress other participants' contributions.

At the same time, there are disadvantages. An asynchronous group may take a long time to respond. Also, contributions may 'come late': discussion may already have moved on, when respondents, after some time, refer to an earlier stage of discussion. The tendency for non-response may be higher than in face-to-face interviews. Technical requirements and problems (e.g. connection difficulties) may influence the data quality, as well as the process of an asynchronous discussion.

Conclusion

The methods presented so far aim at collecting verbal data – whether by interviewing individuals, by inviting them to recount autobiographical experience or by having groups discussing an issue. These methods make knowledge about practices and processes accessible, but do not give immediate access to practices and processes in their course (see also Barbour 2014 and 2018 on the use of focus groups, and Brinkmann and Kvale 2018 on the use of interviews).

Observation

More direct access to practices and processes is provided by the use of observations. Here, we can also distinguish several concepts regarding the role of the researcher:

• Covert and overt observation: how far are the observed persons or fields informed that they are observed?

Whether or not to use covert observation is not so much a technical distinction or linked to the researcher's preferences, as an issue of research ethics (See Chapter 3). Nowadays, such a form of observation without informing participants about the research is hardly justifiable.

- Non-participant and participant observation: how far do observers become active parts of the field that is observed?

Whether or not to use these forms of observation should be linked to the researchers' knowledge interests. In most applications of observation, currently a form of participant observation will be used.

- Systematic and unsystematic observation: is a more or less standardized observation scheme applied or are the processes observed more openly?

Again, it is the research question which should drive the choice between these two alternatives.

- Observation in natural or artificial settings: do the researchers enter the relevant fields or do they 'transfer' the interactions to a specific room (a laboratory) to improve the quality of the observations?

Although much observation-based research in psychology is based on experiments in laboratories (see below), the bulk of using observation in the social sciences relies on applying it in natural settings.

- Introspection or observation of others: in most cases, observation will focus on other people. What role does the researcher's reflecting introspection play in making the interpretations of the observed more solid?

Except for strategies of autoethnography, observation is now mostly directed at observing other people and settings the researchers have not been part of before.

- Standardized and non-standardized observation: in most cases, complex situations are now observed with open and adapted methods. However, approaches using observation categories defined in advance for observing a sample of situations are also applied.

Here, the researchers' knowledge, interests and worldviews drive the decision between these alternatives.

- Experimental and non-experimental observation: in the first case, you will intervene specifically and observe the consequences of such an intervention.

Again, it's the research question and the researchers' worldviews that will drive this decision.

Standardized observations

Bortz and Döring describe this approach as follows:

> The observation plan of a standardized observation exactly defines what to observe and how to protocol what was observed. The events to observe are known in principle and can be dismantled in single elements or segments, which are exclusively the issue of the observer's attention. (2006, p. 270)

For standardized observation, you will draw a sample either of events or of time. The first alternative is oriented to certain events (e.g. a certain activity) in their frequency in the period of observation or in the frequency of occurrence in combination with other events: how often do girls answer in a mathematics class, or how often do they do so after a specific question is asked? To answer these questions, a whole class is observed and the frequency of the events is noted. An alternative is to draw a time sample. Here, the observation is segmented in fixed periods in which you observe or maybe change the object of observation – for example, in five-minute intervals, which are randomly sampled from the observation period (mathematics class). The observations are done with a standardized observation scheme, often after training for the observers in dealing with the situation and the instrument. Observations can be documented with video and then be noted off the tape. A problem can be that the camera (or the observation scheme) does not cover the essentials of the situation. In time sampling, the relevant events may occur outside the selected periods.

Box 11.8 Research in the real world

A student's research study on gender differences in participating in university seminars

In her empirical master's thesis, Julia Asmus (2017) analyzed gender differences in participation in discussions of students in history seminars. She did a standardized observation in six or seven sessions of six seminars and counted each contribution independent of their substance. The process of the discussion in the sessions was protocolled. In addition, she developed and used a questionnaire, filled in by 119 students, concerning the students' motivation to actively contribute to seminar discussions. In relation to the gender distribution among the seminar participants, the observations showed, in 45% of the analyzed seminar sessions, that male students made more frequent contributions. In the questionnaire data, it became evident that male and female students saw the relevance of expressing their own opinions in the seminars quite differently.

This basic situation of social science observation – the researchers observe the field, and the people in it, by using a sample and confine themselves to noting the processes – can be extended in several directions in research practice.

Figure 11.6 visualizes the degrees of standardization in research based on observation.

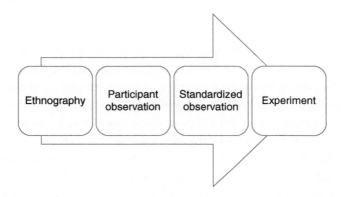

Figure 11.6   Degrees of standardization in observation

Experiments

In experimental research, especially in psychology, observation may be focused on a deliberate intervention in one group, which is then compared to a second group in which this intervention is absent. For example, in school observations, the teacher may apply an intervention, reducing aggression in one group – e.g. a teaching unit about de-escalation in situations of conflict. Then, in a second group, this intervention is not applied. The two groups will then be observed in their behavior in the next situation of conflict that occurs and be compared. If the groups are set up by random sampling, this will be an experimental study. If the groups already exist (two seventh grades), this is a quasi-experimental study. The observation is applied according to the principles mentioned earlier – for example, in sampling times or persons. The researchers will do a non-participant observation and use an observation protocol for documenting behaviors defined as relevant in advance. In our example, the data will still be collected in a field observation (in class). This will make standardization of the research situation more complicated. Therefore, observation in the laboratory, i.e. an artificial situation under controlled conditions, is the alternative to observing in the natural context of teaching situations.

Participant observation

A contrasting form of data collection is provided by participant observation. Here, the researcher's distance from the observed situation is reduced. Their participation over an extended period in the field that is studied becomes an essential instrument of data collection. At the same time, the observation is much less standardized. Here, you will also do some sampling of the situations that are observed, but not in the sense of time sampling as described above. Rather, you will select situations, persons and events according to how far the interesting phenomenon becomes accessible in this selection. The principal procedure can be summarized in Jörgensen's words by 'a logic and process of inquiry that is open-ended, flexible, opportunistic, and requires constant redefinition of what is problematic, based on facts gathered in concrete settings of human existence' (1989, p. 13).

Participant observation can be understood as a two-part process. First, the researchers are supposed to become participants and find access to the field and the persons in it. Second, the observation itself becomes more concrete and more strongly oriented to the essential aspects of the research question. Here we can distinguish three phases (see Spradley 1980, p. 34). First, descriptive observation for orienting in the study field is supposed to provide rather unspecific descriptions, to cover the complexity of the field as far as possible and to make research questions more concrete. Second, focused observation is more and more limited to the processes and problems that are particularly relevant to the research question. Selective observation at the end of data collection is supposed to find further evidence and examples for the processes identified in the second step. The documentation mostly consists of detailed field notes of protocols of situations. Whenever possible, research ethics (see Chapter 3) demand that observations are conducted openly, so that the observed people know that they are being observed and have agreed beforehand to being observed. Third, for (participant) observation, the problem often is that certain issues are not immediately accessible at the level of practice, but only or mainly become 'visible' in interactions when people talk about the issues. Some topics are only an issue in conversations about the research or in ad hoc interviews. However, the results of a participant observation will be more fruitful when more insights come from protocols of activities and fewer come from reports about activities. Nevertheless, conversations, interrogations and other data sources will always comprise a big part of the knowledge process in participant observation.

Ethnography

Recently, the more general strategy of ethnography has tended to replace participant observation. Ethnography, however, is not a subset of observation but has become the broader strategy of today, in which observation and participation are interlinked with other procedures at the same time. In its most characteristic form it involves the ethnographer participating, overtly or covertly, in people's daily lives for an extended period of time, watching what happens, listening to what is said, asking questions – in fact, collecting whatever data are available to throw light on the issues that are the focus of the research (Hammersley and Atkinson 1995, p. 1).

Thus, observation is only one method of many empirical strategies used in ethnography at the same time. This strategy will help you adapt data collection to your research question and to conditions in the field most consistently. Methods are subordinated to the research practice in the field. There is a strong emphasis on exploring a field or phenomenon. You will collect mostly unstructured data instead of using categories defined in advance and an observation scheme. For this purpose, a few cases are involved (or even a single case). Data analysis focuses on interpretation of meanings and functions of practices, statements and processes (see Hammersley and Atkinson 1995, pp. 110–11). Lüders sees the central defining features of ethnography as follows: first [there is] the risk and the moments of the research process which cannot be planned and are situational, coincidental and individual ... Second, the researcher's skillful activity in each situation becomes more important ... Third, ethnography ... transforms into a strategy of research which includes as many options of collecting data as can be imagined and are justifiable. (1995, pp. 320–1; see also Lüders, 2004a)

Virtual Ethnography

If you transfer methods of surveying or individual or group interviews to online research, you turn to the Internet as a *tool* to study people you could not otherwise reach. But you can also see the Internet as a *place* or as a *way of being* (for these three perspectives, see Markham 2004). In these cases, you can study the Internet as a form of milieu or culture in which people develop specific forms of communication or, sometimes, specific identities. This requires a transfer of ethnographic methods to Internet research and to studying the ways of communication and self-presentation on the Internet: 'Reaching understandings of participants' sense of self and of the meanings they give to their online participation requires spending time with participants to observe what they do online as well as what they say they do' (Kendall 1999, p. 62).

The example in Box 11.9 of a case study of doing ethnography online illustrates this.

Box 11.9   Research in the real world

Virtual ethnography

In her study, Hine (2000) took a widely discussed trial (the Louise Woodward case – a British au pair who was tried for the death of a child she was responsible for in Boston) as a starting point. She wanted to find out how this case was constructed on the Internet by analyzing webpages concerned with the issue. She also interviewed web authors by e-mail about their intentions and experiences, and analyzed discussions in newsgroups in which 10 or more interventions referring to the case had been posted. She used www.dejanews.com for finding newsgroups. At this site, all newsgroup postings are stored and can be searched by using keywords. Her search was limited to one month in 1998.

Hine posted a message to several of the newsgroups which had dealt with the issue more intensively. However, the response was rather limited, as other researchers had obviously found repeatedly (2000, p. 79). Hine also set up her own homepage and referred to it when contacting prospective participants or in posting messages about her research. She did this to make herself and her research transparent for potential participants.

In summarizing her results, she stated:

The ethnography constituted by my experiences, my materials and the writings I produce on the topic is definitely incomplete ... In particular, the ethnography is partial in relation to its choice of particular applications of the Internet to study. I set out to study 'the Internet', without having made a specific decision as to which applications I intended to look at in detail. (2000, p. 80)

Nevertheless, Hine produced interesting results on how people dealt with the issue of the trial on the Internet. Her thoughts and discussions on virtual ethnography are very instructive beyond her own study. However, they also show the limitations of transferring ethnography – or, more generally, qualitative research – to online research, as Bryman's critical comment illustrates: 'Studies of online communities invite us to consider the nature of the Internet as a domain for investigation, but they also invite us to consider the nature and the adaptability of our research methods' (2016, p. 448).

In general, interest in doing ethnographies of the Internet – also called 'netnographies' (Kozinets 2010; Kozinets et al. 2014) – is increasing. Often, forms of interviewing are used, as opposed to ethnographic methods in the strict sense of the term.

Box 11.10 Research in the real world

Tasks of online ethnography

Hart (2017, p. 4) discusses four research tasks that online ethnographers engage in and which show the relevance of the research experience for real-world activities: (1) online ethnographers navigate the site, learn about its residents of this virtual world, the different spaces it consists of, and the activities that happen; (2) if they are new to the environment, online ethnographers learn how to become competent community members; (3) online ethnographers build and maintain relationships with research participants, often aiming at identifying local informants; and (4) they should carefully and consistently document what they see, do and learn during their (participant) observations, which means they must actively collect data.

Conclusion

The observational methods described above vary in the distance the researchers maintain to the field that is observed; the alternatives are to participate or just to observe from the outside. Furthermore, the methods differ in the degree of control of conditions of the study exerted by the researchers. Control is strongest in the laboratory experiment and weakest in participant observation. The methods can also be distinguished by the standardization of the research situation – again, most limited in participant observation and strongest for the experiment. In general, for observation, the idea guiding data collection is that it provides more immediate access to practices and routines compared to interviews and surveys. However, in most cases, conversations, statements and questions, or sometimes ad hoc interviews, are involved in observations.

Obtaining and Documenting Information

Social research is based on data collected with empirical methods. In general, we can distinguish two major groups of methods. Quantitative methods aim at covering the phenomena under study in their frequencies or distributions and therefore work with big numbers of cases in data collection. Numbers are in the foreground. For example, you could tally how often waiting times in the hospital occur, what the average time is that a patient waits (unnecessarily) before a treatment and how this is distributed over the week.

Qualitative methods are more interested in an exact description of processes and views and therefore often work with small numbers of cases. In the foreground are texts, such as transcribed interviews. In our example, you would collect data about how situations occur in which patients wait (unnecessarily), what subjective explanations the staff have for this or how particular patients experience these situations of waiting.

Data collection in qualitative research pursues different aims and is grounded in different principles, from quantitative research. Quantitative research is devoted to the ideals of measurement and works with numbers, scales and index construction. Qualitative research is more oriented to producing protocols of its research issues and to documenting and reconstructing them. Before we turn to methods of analyzing data (see Chapter 12), we will briefly consider these aims and principles.

Measurement

For quantitative methods, a measurement is assumed; in a time measure, for example, the duration of an event is identified with a measurement instrument (a watch). This is fairly unproblematic if there is an established unit (e.g. a minute, a centimetre) and a means of measurement to identify how many of these units are given in the concrete case (e.g. 15 minutes of waiting time). However, often this unit does not exist for the objects social science is interested in and has to be defined by the researcher. To measure then is to allocate a number to a certain object or event. Three problems are linked to this allocation. First, the number comes to represent the object or its feature in the further progress of the research – in other words, the object itself is no longer part of the process. In addition, different numerical values represent differences and relations between the objects. For example, extensions of waiting time before an operation are measured in minutes. In this case, you can assume that two minutes are always two minutes and four minutes are twice as long as two minutes. Another example illustrates the other two problems, namely unambiguousness and significance. The subjective distress of the patient has to be translated into a numerical value first. If 'extremely stressful' equals 4, 'very stressful' equals 3 and 'stressful' equals 2 on a scale, you can assume neither that the distance between 2 and 3 is as big as that between 3 and 4, nor that a value of 2 always represents the same degree of subjective distress. This is the question of the unambiguousness of measurement values. Finally, you cannot assume in this example that a value of 4 here represents double the value of 2. This is the problem of significance, which raises questions about which mathematical operations make sense based on the measurements made.

Scaling

The allocation of numerical values to an object or event leads to the construction of a scale. Here, four kinds of scales are distinguished.

A *nominal scale* allocates objects with identical features to identical numerical values. For example, male as gender might be labeled with 1, female as gender with 2. A relation between the values does not exist.

In *ordinal scales*, a relation of ordering exists between the values. If, in our example, the degree of the patients' subjective distress is labeled as 4 when the situation is extremely stressful, 3 when it is very stressful and 2 when it is stressful, this represents an order between the different degrees of subjective distress. But here, the distances between the values are not necessarily the same.

In an *interval scale*, in contrast, the distances between two values are always the same. An example is the Fahrenheit temperature scale.

A *ratio scale* is given if the distances between values are the same, but also if you can assume that two units of distance represent twice the distance of one unit. Examples are measurements of length or weight: the distance between 2 kg and 3 kg is the same as the distance between 5 kg and 6 kg, and 6 kg is twice as heavy as 3 kg. A ratio scale furthermore has a fixed point of zero. Ratio scales are seldom found in the social sciences and their research.

The kind of scale determines which calculations are justified for each scale (see Chapter 12). Table 11.1 summarizes the various kinds of scale.

Counting

If you want to count certain objects, this assumes that the objects to be counted are equal in their major features: that is, that you are not 'comparing apples and oranges'. If you want to count people, activities or situations, the sameness as a precondition of the countability has to be produced in advance. People are characterized by strong individuality and diversity. They can be counted if they can be classified according to specific features, such as their age. Age is already defined as a number. For other, more qualitative features, a classification in numerical form first has to be done. Someone is either male or female. People can be helpful to varying degrees (not at all, a bit, averagely, very or extraordinarily helpful). To be able to count such a feature, the characteristic has to be classified and labeled with a numerical value ('not at all helpful' is classified as 1, 'a bit helpful' as 2, etc.). Then you can count these features and relate them to others using numerical values. A precondition is that a feature like helpfulness can be classified by such categories. For this purpose, it must be possible to define categories exactly (e.g. what does 'a bit helpful' or 'not at all helpful' mean?). Categories should be able to describe the feature in an exhaustive way (it must be possible to allocate all possible forms of helpfulness to these categories).

Table 11.1 Kinds of scale

Kind of scale	Possible statements	Examples	Answers and values	
Nominal scale	Equality differences	Gender	☐	Male
			☐	Female
		Professional groups	☐	Physician
			☐	Teacher
		Satisfied by the treatment	☐	Yes
			☐	No
Ordinal scale	Bigger–smaller relations	Social status	☐	Upper class
			☐	Middle class
			☐	Working class
		School grades	☐	Very good
			☐	Good
			☐	Satisfactory
			☐	Sufficient
			☐	Failed
Interval scale	Sameness of differences	Temperature (°F)	☐	36 °F
			☐	37 °F
			☐	38 °F
		Calendar, time intervals (e.g. of sickness leave per year)	☐	One day
			☐	Two days
			☐	Three days
Ratio scale	Sameness of relations	Weight	☐	1 kg
			☐	2 kg
			☐	3 kg
		Length	☐	1 cm
			☐	2 cm
			☐	3 cm

Constructing an index

For quantitative features like age, all persons with a certain value (age) are summarized (e.g. all 25-year-olds) or allocated to specific age groups (e.g. the 20- to 30-year-olds). Often, more complex features have to be counted. Many studies start from the social status of a person (e.g. to find out whether people with a high social status fall ill less often than people with a low social status). Social status is constructed from various single features – level of education, profession, income, housing situation. This means that an index is constructed in which these single features are combined. Counting is then applied to this index. The parts of the index can have the same weight or can be weighted differently, such as when income counts for twice as much as the other parts of the social status. In the research on quality of life (see, for example, Guggenmoos-Holzmann et al. 1995), a variety of quality of life indices have been established.

Protocols

While quantitative research is based on measurement and counting, qualitative research tends to refrain from using such numerical values. Rather, the first step is to produce a protocol of the events and of the context in which they occurred. The protocol should be as detailed, comprehensive and exact as possible. For observations, you will produce detailed descriptions of situations and of their contexts. Interviews are recorded on tape or on mp3 recorders, and this is complemented by memory protocols of the interview situation. Interactions are documented on audio or video tape in order to make possible repeated and more or less unfiltered access to the raw data. Whereas in interviews the data are produced with methods (questions lead to answers or narratives, which are produced specifically for the research), several other approaches in qualitative research, like ethnography or conversation analysis (see Chapter 12 and Flick 2014, 2018a for more details), restrict themselves to recording and protocolling everyday life situations, without intervening with questions, for example. This is also discussed as using 'naturally occurring data' (Potter and Shaw 2018). Quantitative research produces a specific condensation of the data already in the data collection by delimiting them to specific questions and possible answers. Qualitative research is first interested in less condensed data. Reduction of the information here is part of the analysis. The original data should remain available in an unfiltered way and accessible repeatedly.

Documentation

Documentation of the data has a specific relevance here. A comprehensive recording, which is then turned into a transcript (see also Chapter 12), is seen as most important for qualitative research with interviews or focus groups. Therefore, audio or video recording has priority over making notes of answers or practices only at a glance. Only where technical recording devices are in the way of the method – when they prevent the researcher from integrating into the field in participant observation, for example – is the preference still for field notes and protocols made after the observation (see Flick 2018a, Chapter 24 for more details). A detailed and comprehensive recording leads in most cases to a similarly exact transcript of the data. This should include as much as possible of the context information for the interviewees' statements. Here again, an unobstructed view of the reality under study is seen as highly relevant.

Exactness in documenting the events is a precondition for a detailed interpretation of the statements and occurrences grounded in the data.

Reconstruction

In the analysis and interpretation of the data, it becomes possible to reconstruct it – asking, for example, how something does or did occur, or what views the participants had of this occurrence. In a detailed reconstruction of case trajectories, data are obtained which will then be the subject of comparison. This comparison does not work with numbers, but nevertheless is intended to arrive at generalizing statements. For example, reconstructions of single trajectories lead to constructing types of processes and detailed descriptions of these types. Here, it is less important how often these types can be identified. Rather, the question is how far these types cover the range of existing trajectories and allow inferences about when and under what conditions each type is relevant.

Summary

The two kinds of methods – quantitative and qualitative – are often characterized by their essential difference in the standardization of the procedures. Analyzing data from a questionnaire in a quantitative and statistical way makes sense only when the data collection is standardized by uniform question wording and sequence, and uniform alternatives for answering them. This requires that every participant is under exactly the same conditions in answering the questions. Qualitative studies in contrast are often most fruitful when the procedures are less standardized and are applied in a flexible way so that new and unexpected aspects become relevant. At the same time, specific attention is paid to including context information: the participants' answers are not meant to speak as facts for themselves. Rather, they should be embedded in a longer narrative or extended presentation. This will then allow insights into the subjective meaning of what has been presented and thus make clear in which contexts the interviewees themselves understand their statements.

What You Need to Ask Yourself

In doing your empirical project in social research, you should consider the points in Box 11.11 for selecting and conceptualizing methods of data collection. These questions can be helpful in doing your own study and also in assessing the studies of other researchers.

Box 11.11 What you need to ask yourself

Data collection

1 What are the major aspects that should be covered by the data collection?
2 Is the focus more on practices referring to an issue or more on knowledge about it?
3 Do existing data already include the relevant information, so that it is not necessary to collect your own data?
4 Do the existing data and their content, their degree of detail, fit your own research question, or are they different enough to justify your own, new research?
5 What is the plan for the data in their later analysis? Are the data provided by the actual method of data collection appropriate for this kind of analysis (in their exactness, structure and level of scaling)?
6 What scope for idiosyncrasies (of the contents or the way of presenting them by the participants) do the selected methods offer, and what scope is necessary to answer the research question?
7 What degree of exactness in documenting and transcribing qualitative data is necessary to answer the research question?

What You Need to Succeed

The questions in Box 11.12 give an orientation for doing your data collection.

Box 11.12 What you need to succeed

Data collection

1 What are the major aspects that should be covered by the data collection?
2 What characterizes the method you want to use for your data collection?
3 If necessary, have you read additional resources about how to use this method and about other researchers' experience of it?
4 Is the method you selected producing the data you expect and need?

What you have learned

- Surveys, interviews, observation and use of existing data are the major methods in social research.
- These methods play a part in both qualitative and quantitative research.
- Surveys can be done in an open way or with several predefined possibilities for answering the questions.
- They can be realized with questions or narrative stimuli, with single participants or in groups.
- Observation can be applied in a standardized way or can be open and participant-based.
- Secondary or material analyses have to take the inherent structure of their data into account.
- In quantitative research, data are collected in measurements and on different levels of scaling, which then have implications for the types of analysis made possible.
- In some cases, events are counted. Sometimes features can be accessed only indirectly through the construction of an index covering several particular features.

What's next

The first and fourth texts listed below provide a comprehensive overview of social research methods with a stronger focus on quantitative methods, while the second and third concentrate more on qualitative methods:

Bryman, A. (2016) *Social Research Methods*, 5th edn. Oxford: Oxford University Press.
Flick, U. (forthcoming) *Doing Interview Research*. London: Sage.
Flick, U. (2018) *The SAGE Handbook of Qualitative Data Collection*. London: Sage.
Neuman, W.L. (2015) *Social Research Methods: Qualitative and Quantitative Approaches*, 7th edn. Boston: Allyn & Bacon.

The following two articles outline online ethnography as a method, as an example:

Hart, T. (2014) 'Technologies for Conducting an Online Ethnography of Communication: The Case of Eloqi', in S. Hai-Jew (ed.) *Enhancing Qualitative and Mixed Methods Research with Technology*. Hershey, IGI Global. pp. 105–24.
Hart, T. (2017) 'Online Ethnography', in J. Matthes, C. S. Davis and R. F. Potter (eds) *The International Encyclopedia of Communication Research Methods*. Hoboken, NJ: Wiley.

The following case study can be found in the online resources. It should give you ideas about how several of the methods discussed in this chapter can be used in a research project about students' Facebook use:

Stirling, E. (2014) 'Using Facebook as a research site and research tool', *SAGE Research Methods Cases*. doi: 10.4135/978144627305013510242.

Research Methodology Navigator

Orientation

- Why social research?
- Worldviews in social research
- Ethical issues in social research
- From research idea to research question

Planning and design

- Reading and reviewing the literature
- Steps in the research process
- Designing social research

Method selection

- Deciding on your methods
- Triangulation and mixed methods

You are here in your project

Working with data

- Using existing data
- Collecting new data
- Analyzing data

Reflection and writing

- What is good research? Evaluating your research project
- Writing up research and using results

12

ANALYZING DATA

How this chapter will help you

You will:

- understand the logic of analyzing data in social research
- know some major methods of data analysis in social research
- understand the similarities and differences between the procedures of qualitative and quantitative data analysis, and
- assess which methods are available for your research and what they can provide.

The previous chapters discussed selected methods of using existing data (Chapter 10) or for collecting new data (Chapter 11). In this chapter, we turn to describing methods for analyzing the data that have been selected or collected. We first consider quantifying analyses for standardized data. Then we look at interpretative methods for analyzing qualitative data from interviews and participant observations. The process of any data analysis includes several steps – data have to be:

1 elaborated on for the actual analysis;
2 checked for their consistency (e.g. by eliminating contradictions or unclear parts);
3 regarded for what to find out from analyzing them;
4 processed by applying the methods of analysis;

5 interpreted for the meaning of the findings they offer;

6 focused on the major findings and conclusions they allow;

7 evaluated for their quality and that of the analysis (see Chapter 13); and

8 turned into a presentation of the knowledge that was developed from them (see Chapter 14).

Quantitative Data Analysis

In the first part of this chapter, we will address a number of basics in quantitative data analysis. Before we turn to the analysis of survey data, we will first consider quantitative content analysis. In the second part of the chapter, we will have a look at some qualitative approaches to analyzing data, including qualitative content analysis.

Quantitative Content Analysis

Content analysis is a classical procedure for analyzing textual material of whatever origin, from media products to interview data. It is 'an empirical method for systematic, inter-subjectively transparent description of substantial and formal features of messages' (Früh 1991, p. 25). The method is based on using categories derived from theoretical models. One normally applies such categories to texts, rather than develop them from the material itself – though one may of course revise the categories in the light of the texts under analysis. Content analysis aims at classifying the content of texts by allocating statements, sentences or words to a system of categories. This approach distinguishes between quantitative and qualitative content analyses, which we will address later.

Analyzing newspaper articles

While *qualitative* content analysis is seen as a method of *analyzing* data from interviews (see below), for example, some sources see *quantitative* content analysis rather as a specific method for *collecting* data (e.g. Bortz and Döring 2006; Bryman 2016). Schnell et al. (2008, p. 407) see the method as 'a mixture of "analytic technique" and data collection procedure'. It is used for collecting and classifying information, such as in newspaper articles. Bortz and Döring provide the following as a definition: 'Quantitative content analysis captures particular features of text by categorizing parts of the text into categories, which are operationalizations of the interesting features. The frequencies of the single categories inform features of the analyzed text' (2006, p. 149).

In the foreground of such analyses are the questions: (a) what characterizes the communication about a specific issue in certain media, and (b) what impact does this have on the addressees? Communication according to this model can be defined according to Laswell's (1938) formula: *Who* (communicator) *says* (writes, mentions in the form of signs) *what* (message) in *which channel* (medium) to *whom* (receiver) and *with what effect*?

The methodological core of content analysis is the category system used to classify the materials you study. The allocation of a passage in the text to a category is described as coding. A crucial step is that you select the right materials (the sample that is drawn from the text) and the correct units for your analysis. Which texts from which newspaper and which publishing dates should you select? Will you analyze particular words or will you allocate whole sentences or paragraphs to the categories?

Analytic strategies

We can distinguish several analytic strategies. In simple frequency analyses, you ask how often certain concepts are mentioned in the texts you analyze. This method is used to infer the medial presence of a topic in the daily newspapers – for example, how often was the topic 'health fund' an issue in the most important German press releases in the period under study (see Table 12.1)?

Table 12.1   Frequency analysis in content analysis

	Health fund
Newspaper A 20.10.2018	
Newspaper B 20.10.2018	
Newspaper A 21.10.2018	
Newspaper B 21.10.2018	
Newspaper A 22.10.2018	
Newspaper B 22.10.2018	
Newspaper A 23.10.2018	
Newspaper B 23.10.2018	

In a contingency analysis, you do not seek only the frequency of concepts (and thus of topics) in the relevant period in the press. Contingency analysis is interested in which other concepts appear at the same time – for example, how often is 'health fund' mentioned together with 'caring deficits' or 'costs' (see Table 12.2)?

Table 12.2 Contingency analysis in content analysis

	Health fund	Caring deficits	Costs	Interest of health insurers
Newspaper A 20.10.2018				
Newspaper B 20.10.2018				
Newspaper A 21.10.2018				
Newspaper B 21.10.2018				
Newspaper A 22.10.2018				
Newspaper B 22.10.2018				
Newspaper A 23.10.2018				
Newspaper B 23.10.2018				

Quantitative content analysis is often used to analyze newspaper articles. However, you can use it as a method for analyzing interviews or other materials that have been produced for research purposes.

Steps in quantitative content analysis

We can identify several steps in the analysis. First, you will decide which texts are relevant for the purpose of your study. The next step is to draw a sample from these texts before you define the counting unit (all or certain words, groups of words, sentences, complete articles, headlines, etc.). From the research question and from its theoretical background, you will next derive a system of categories. These should be (a) mutually exclusive (clearly distinguishable), (b) exhaustive, (c) precise, (d) based on discrete dimensions, and (e) independent of each other.

The classification of texts into categories aims mainly at reducing the material. The category system can consist of concepts and subconcepts. Often, a dictionary is established that includes the names of categories and the definitions and rules for allocating words to categories. Coding rules are defined so that categories can be applied to texts. Coders collect the analytic units by applying the categories that were defined before. Coders are trained for this purpose and the coding system is checked in a pre-test for its reliability (correspondence of allocation by different coders). Then statistical analyses are applied to identify how often certain words in total or in connection with other words appear in the text and to analyze the distribution of categories and contents (see below).

Problems of quantitative content analysis

Problems in quantitative content analysis arise from the necessary isolation of single words or passages, which are thus taken out of context. Texts are decomposed into their elements, which can then be used as empirical units. This makes it more difficult to find any meaning or coherence in texts.

For example, there have been attempts to identify changes in attitudes towards the elderly by analyzing newspaper articles using the frequencies with which certain words (e.g. 'frail' or 'experience') appear together with 'age' or 'old people'. If the concepts or frequencies with which they are used together with 'age' and 'old people' are changing, this is used to infer changes in attitudes towards aging.

Thus, the application of content analysis is often rather reductionist. This may result from strong standardization and the use of small analytical units (e.g. a particular word) in order to provide repeatability, stability and exactness of the analysis. Repeatability refers to the degree to which classifications of the material are completed in the same way by several analysts. Stability means that the method of classification of content does not vary over time. Exactness indicates how far the coding of a text corresponds with the norm of coding or the standard coding.

A strength of quantitative content analysis is that you can analyze large amounts of data with it. The procedures can be standardized to a high degree. Frequencies and distributions of statements, attitudes, etc. can be calculated. The weakness of quantitative content analysis is that one rules out the analysis of particular cases right from the beginning: the single text and its structure or particularity as a whole are not taken into account. The context of words is rather neglected. How far the analysis of frequencies or topics in texts is sufficient for answering substantive research questions has been debated since the early days of research with content analysis (see also Krippendorf 2012).

Quantitative Analysis of Other Forms of Data

In the first part of this chapter, quantitative approaches to content analysis were presented. In the next step, we will consider general aspects of quantitative data analysis of questionnaire data, for example, before I then turn to qualitative interpretative analyses.

Elaboration of the data

Before you can analyze questionnaire data, you have first to elaborate on them. This includes constructing a data matrix, i.e. a compilation of all variables for every study unit, more specifically of all responses for every case (see Table 12.3), which you will transform into numerical values (see Table 12.4). The questionnaire in our example begins with four questions about the demographic characteristics of the respondents (gender, age, profession, school qualification) before substantive questions follow. A code plan was developed in advance, showing which number is code for which possible answer. For gender, female is coded with 1, male with 2. School qualification is coded from 'without' to 'high school' with values of 1–4, and current profession in a similar way. For the answers to questions 1 and 2, the values are taken from the scale (see the example in Figure 1.1 in Chapter 1). The cases are given an identification number and the variables are labeled with numbers (e.g. age in Table 12.3 becomes variable V2 in Table 12.4).

In this context, coding means allocating numerical values to answers. In this data matrix, you will enter all responses from every questionnaire. If open questions (without a defined scale of answers) are used, the answers (the text noted by the participant at this point) have to be allocated to categories, which can then be labeled with numerical values.

Table 12.3 Data matrix 1

Study unit	Variable					
	Gender	Age	Profession	School qualification	Question 1: grade of consent	Question 2: grade of consent
Case 1	M	21	Student	High school	5	3
Case 2	F	28	Sales person	Grammar school	3	4
Case 3	M	–	Taxi driver	Public school	1	1
Case 4	F	25	Physician	Without	2	5
...						

Table 12.4 Data matrix 2

Study unit	Variable					
	V1	V2	V3	V4	V5	V6
01	2	21	4	4	5	3
02	1	28	1	2	3	4
03	2	999	3	3	1	1
04	1	25	5	1	2	5
...						

Checking for consistency and cleaning the data

In the next step, one needs to check the data for consistency and to clean the data. First, you should test in the first frequency calculation whether data have been entered in the wrong column. If, for example, only 10 possible values were defined for 'profession' but several cases have values of 25 or 35, this indicates that maybe age values were coded in the column for profession. Such mistakes in columns have to be checked for and corrected. Then, missing data have to be checked. In Table 12.3, for case 3, the age is missing and was coded with '999' for a missing value in Table 12.4. You should also do a plausibility check for the data. In Table 12.3, you will find entries in case 4, which are at least unlikely (a physician without a school qualification and with an age of 25). Here, you should also check whether this is a coding error or if a combination of these answers really can be found in the questionnaires. Perhaps these answers have to be treated as 'missing'. After this assessment of the data, which can be quite time-consuming for big datasets, and after correcting all identified errors, you can analyze the data on various levels of complexity.

Clarifying what the analysis will reveal

The next step is to clarify (again maybe) what the analysis of the data will reveal – the guiding research question for the analysis. This will define which kind of analysis you should do and what it should focus on. This will also define which calculations you should apply and which variables you should put in relation to each other.

Univariate analyses: referring to one variable

A common way of demonstrating commonalities and differences for a certain feature or for a variable is to calculate its distribution in the sample that is studied. For this purpose, answers can be analyzed for their frequency, in terms of how often they were given, or for their relative frequency, by dividing the number of cases in one category by the number of cases in the sample.

Frequencies

If there are, say, four possible answers, you can first calculate their relative frequencies by dividing the number of cases in one category by the number of cases in the sample. If you want to calculate the percentage of the frequency, the result of this division is multiplied by 100. If, say, 27 of a sample of 100 people ticked 'public school' as highest qualification, the relative frequency of public school as school qualification is 0.27 and the percentage is 27%. Finally, you can calculate the cumulated relative percentage. If in our example another 33 people indicated 'grammar school', 20 people 'no qualification' and 20 people 'high school', you can rank order the values according to the level of qualification. From cumulating (or adding) the single values, you can see, for example, that here the relative frequency of people with no more than a grammar school qualification is 0.80 and the relative percentage is 80% (see Table 12.5).

To demonstrate the distribution of answers in a sample, you can go two ways: on the one hand you can identify the central tendency, or on the other hand the dispersion.

Table 12.5   Frequency distribution of the variable 'school qualification'

	Category	Number of cases	Relative frequency	Percentage	Cumulated relative frequency	Cumulated percentage
No qualification	1	20	0.20	20%	0.20	20%
Public school	2	27	0.27	27%	0.47	47%
Grammar school	3	33	0.33	33%	0.80	80%
High school	4	20	0.20	20%	1.00	100%

Central tendency

The most prominent measure for the central tendency is the arithmetic mean, which is calculated by dividing the sum of the observed values by the number of cases. A familiar example is school grades. A student has the grades 1, 1, 4, 3 and 5 in his qualification. To calculate the mean, you will add the single grades and divide that by the number of subjects (the sum is 14, the number of subjects is 5, which makes a mean of 2.8). To calculate means, you need data on the level of interval scales (the distances between the values have to be equal – see Chapter 11). If your data only consist of an ordinal scale, you can calculate the central tendency with the mode or with the median.

The mode is the value that occurs most frequently. The mode in our grades example is 3 (as it occurs most often). The median is the midpoint of a distribution, which means where the cumulated relative frequency reaches 50%. This means the distribution is separated such that 50% of the values are under and 50% are over the median. In our example of school qualifications (see Table 12.5), the median would be a little higher than 'public school' as almost 50% of the respondents have the values 'public school and less' and a little more than 50% have 'grammar school and more'.

Dispersion

Measures of central tendency will not tell you everything about a distribution. The example of the mean of two school grades may demonstrate this. A grade of 3 and a grade of 4 have a mean of 3.5. The same applies to the grades of 1 and 6. The dispersion of the grades is much lower in the first case (both values are close to each other) than in the second case (where the distance between the values is much bigger). To take this dispersion into account, the first way is to calculate the range of the values. This is the difference between the minimum and the maximum value. For this purpose, you will subtract the minimum from the maximum values. In our example, in the first case (grades of 3 and 4), the range is 1. In the other case (grades of 1 and 6), it is 5.

The range is still strongly influenced by outliers, as it is only based on minimum and maximum values; it does not take the frequency of the values between those outliers into account. By defining the quartiles and the distance between quartiles, you can analyze the distribution of the values more exactly. Quartiles define the limiting points which distinguish the quarters of a value distribution. Accordingly, there are the quartiles Q_1–Q_4: 25% of the values are less than or equal to the value of the first quartile Q_1, 50% are less than or equal to the second quartile Q_2, and 75% are less than or equal to the third quartile Q_3. The second quartile is equivalent to the median and separates the second and third quarters of the values. The interquartile range is the difference between the third and first quartiles.

If the measured values are expressed on an interval or a ratio scale, you can calculate the dispersion by calculating the *standard deviation* and the *variance*. The standard deviation is the average amount of variation around the mean and the variance is the squared value of the standard deviation. These measures shed light on the distribution of the single values in the sample. With the calculations discussed so far, you can identify the central tendency in the data, what the average values are in a variable and how the values are distributed in the dataset.

Analyses referring to two variables: correlations and bivariate analyses

If you want to identify the connections between two variables, you may calculate their correlation. Correlation means that a change in the value of one variable is associated with a change in the other variable. Three forms of correlation can be distinguished: a positive correlation (when variable 1 has a high value, variable 2 also has a high value); a negative correlation (when variable 1 has a high value, variable 2 has a low value and vice versa); and the absence of a correlation (you cannot say what will be the value of variable 2 whether the value of variable 1 is high or low). The correlation coefficient varies between –1 (a strongly negative correlation) and +1 (a strongly positive correlation), while a value of 0 indicates the absence of a correlation. For example, you will find a correlation between education and income (more education is associated with higher income).

Correlations need to be interpreted: a correlation does not by itself indicate which of the variables is causal (for example, does more education lead to more income, or does a higher income make more education more likely?). Furthermore, a correlation does not establish a causal connection. In our example, perhaps another variable (e.g. the social status of the family of origin) is the reason for the high values for education and income: the two are consequences of the third variable and are not in a causal relation with each other. On the level of calculation, you may find meaningless connections in calculations (e.g. links between good weather and income). Often, the issues under study are more complex than can be represented through bilateral correlation, which means that both variables are distinguished for two values and related to each other. This approach can be extended to analyses with more than two variables (multivariate analyses).

Methods of multivariate analysis can also be used to test relations, especially differences between groups, if the data are on interval or ratio levels. Multiple regression, for example, starts by analyzing the differences between means in groups and shows how far a set of variables explains the dependent variable, in the sense of predicting values of the dependent variable on the grounds of information about independent variables. In addition, the regression measures the direction and strength of the effect of each variable on a dependent variable (see Neuman 2014, p. 421).

This kind of analysis can not be discussed here in more detail (but see Neuman 2014 and Bryman 2016 for more details). Software packages like SPSS facilitate such multivariate analyses. Nevertheless, the following should be borne in mind:

> It is typical for … multivariate methods that they hardly produce unambiguous solutions …. That leads to the general problems that these procedures now can be used quite easily with the current software packages for statistical analyses. However the question of which model to choose and the interpretation of the results that are produced still require firm methodological skills and refined theoretical considerations referring to the issue of the studies. (Weischer 2007, p. 392)

Testing associations and differences

To find relationships between variables in the data is often not sufficient for answering a research question. Rather, it becomes necessary to test whether the observed relationships occur by chance and how strong the observed relationships between two variables are. It might also be necessary to know whether one variable is the cause of the other or if both are mutual conditions for each other. To answer such questions, you can apply a significance test. Then you basically test whether a result was to be expected or whether it represents a relationship between two variables that is beyond chance.

If, say, you want to study pupil absence from school because of illness and to find out whether there is a relationship to gender, you can count the frequencies of absence for boys and girls. If you find that 76% of the documented absences occur for female students, this may be remarkable on first sight and seem to indicate a relationship between gender and absence. However, if in the school that is studied and in the sample that was drawn three-quarters of the students are girls, it can be expected that 75% of the absences will also apply to girls. That means that you should test whether there is enough difference between the measured value (in our example 76%) and the expected value (in our example 75%) that you can derive a relationship between the variables – here, gender and periods of absence. By comparing the expected and the observed values, you will test the null hypothesis: there is no relation between the variables beyond chance. If in our example the value of 76% has been observed, you can test in a significance test (e.g. the chi-square test) whether the difference between expected and observed values is big enough to confirm a relation between gender and absence. For this purpose, several significance tests are available that can be applied depending on the type of data that were measured. These include the *t*-test and the Mann–Whitney test.

In the *t*-test, two datasets are compared for their differences – for example, measures at two times or measures of two subgroups, such as either (a) the grades of a school class at the beginning and at the end of a term or (b) the grades of two classes at the end of term. With the *t*-test, a statistical significance test is applied, in which the means and the standard deviations of both datasets are taken to calculate the probability that the differences between the two datasets are accidental. Here, you will also assume that the null hypothesis is correct – that there are no real differences and that observed differences occur by chance – until the statistical test demonstrates that the probability is big enough that the differences found are beyond chance. The latter is the case when you can show that there is a probability of less than 5% that differences are accidental. The *t*-test can be applied to small samples and also to samples with different sizes – like school classes of 18 and 25 students in our example. This test requires interval data.

The Mann–Whitney *U*-test can be applied to ordinal data. Here, you will do a rank ordering of the cases in both groups and an overall ranking of all cases in both groups. The position of the members of each group in this overall ranking will be the basis for calculating and testing the statistical significance of the differences between the two groups.

Descriptive and inferential statistics

A brief overview of quantitative analysis focuses on the basics of descriptive and inferential statistics. The latter kind of analysis is mainly relevant for quantitative research, showing relations and patterns in the sample that has been studied to make inferences about the basic population it was drawn from. Descriptive statistics focus more on describing the features of the data in a sample (e.g. gender or age distribution). Calculating central tendencies (e.g. the mean of the ages of participants in a study) will only be the first step in a quantitative study that is complemented by inferential analysis. However, such descriptive statistics can give some context information in a qualitative study, too, for locating a particular case that is analyzed in the field or the range of participants in the study (see also Chapter 9 on combining both approaches).

Quantitative analyses of relations: a conclusion

The procedures for analyzing relations outlined above highlight either the central tendency in the data or the distribution of values in the data. They also calculate the relations between two or more variables and test the significance of the relations that were found. As Worthington and Holloway (1997, p. 76) have shown, the variable types define which kinds of calculations are adequate. For nominal scale variables, the central tendency is measured by the mode and the dispersion is analyzed with the distributions of frequencies. Adequate statistical tests can be based on Chi-square tests. With ordinal scale variables, you can calculate the median for the central tendency and the range for dispersion, and the Mann–Whitney U-test can be applied. For interval scale variables, you can measure the central tendency and the dispersion with the standard deviation, and use the t-test.

Our short overview has presented only some basic principles of quantitative analyses. To apply and study them more extensively, you should consult a more comprehensive textbook on quantitative research and statistics (e.g. Neuman 2014; Bryman 2016). Relations between variables that might be found or tested always have to be interpreted for their meaning in answering the research questions. The example of meaningless connections in calculations of correlation has been mentioned above already. Thus, the process from data to analysis by using methods and software packages only leads to an end, if the major results of the (statistical) analysis can be interpreted in a meaningful way. Finally, the question has to be answered whether the findings can be generalized – in a typology, or a pattern, for example, and beyond the sample that was studied (see also Chapter 13 for the limits of quantitative analysis).

Qualitative Analysis

The approaches discussed so far have focused on quantitative data and quantitative analysis – the frequency and distribution of features in newspaper texts and questionnaires, for example. We will now address some approaches of qualitative analysis, beginning with qualitative content analysis and turning to more interpretative approaches later.

The analysis of qualitative data can be oriented to various aims. The first aim is to *describe* a phenomenon, for example the subjective experiences of a specific individual or group (e.g. the way people carry on with their lives after a terminal diagnosis). Such a study could focus on a case (an individual or a group) and its special features and links between them. Or the analysis could focus on *comparing* several cases (individuals or groups) and on what they have in common, or on the differences between them. The second aim is to identify the conditions on which such differences are based – which means looking for *explanations* for such differences (e.g. circumstances which make the coping strategy for a specific illness situation more successful than in other cases). The third aim is to *develop a theory* of the phenomenon under study from the analysis of empirical material (e.g. a theory of illness trajectories).

Our brief model of the process for data analysis, presented at the beginning of this chapter, roughly applies to qualitative analysis as well and includes the following steps. Data should be:

1 elaborated on for the actual analysis, for example by transcription
2 checked for their consistency (e.g. by clarifying unclear parts)
3 regarded for what to find out from analyzing them by defining a research question for the analysis
4 processed by applying the methods of analysis, for example coding procedures
5 interpreted for the meaning of the findings they offer
6 focused on the major findings and conclusions they allow, for example by developing a typology
7 evaluated for their quality and for that of the analysis (see Chapter 13), and
8 turned into a presentation of the knowledge that was developed from them (see Chapter 14).

Transcription of Interview Data

Interviews are almost always recorded with a mp3-recorder, a smartphone or sometimes a video camera. Transcription is the step in which the information (statements, narratives, etc.) given and recorded in an interview is turned into data that can be analyzed. Rules regarding how to transcribe are used to produce the data in a comparable way, independent of who exactly does the transcription.

Rules for transcription

One example of rather simple transcription rules is given in Box 12.1.

Box 12.1   Research in the real world

Rules for transcription

Layout

Margin	Left 2, right 5
Line numbers	running through or every page starts again
Lines	1,5
Page numbers	On top, right
Interviewer:	I: Interviewer
Interviewee:	IP: Interviewee

Transcription

Spelling	Conventional
Interpunctuation	Conventional
Breaks	Short break: *; more than 1 sec *no. of seconds*
Incomprehensible	((incomp))
Uncertain transcription	(abc)
Loud	With commentary
Low	With commentary
Emphasis	With commentary
Break off word	Abc-
Break off sentence	Abc-
Simultaneous talk	#abc#
Paralinguistic utterance	With commentary (e.g. sighs …)
Commentary	With commentary
Verbal quote	Conventional
Abbreviations	Conventional
Anonymization	Names with°

If you use the suggestions in Box 12.1 to transcribe your interviews, transcripts like that in Box 12.2 should result.

Box 12.2 Research in the real world

An example from a transcript

1	I:	Yeah the first question is, what is this for you, health? ((telephone rings)) Do you
2		want to pick it up first?
3	N:	No.
4	I:	No? Okay.
5	N:	Health is relative, I think. Someone can be healthy, too, who is old and has a
6		handicap and can feel healthy nevertheless. Well, in earlier times, before I came to
7		work in the community, I always said, 'someone is healthy if he lives in a very well
8		ordered household, where everything is correct and super exact, and I would
9		like to say, absolutely clean'. But I learnt better, when I started to work in the
10		community (…). I was a nurse in the NAME OF THE HOSPITAL-1 before that, in
11		intensive care and arrived here with …

KEY: I = Interviewer; N = Nurse

Checking for consistency and completeness of the data

Transcription is the form used to elaborate on the data in qualitative research. Transcripts have to be checked for their completeness, or for the completeness necessary to do the study and the analysis that is planned. Consistency means that all transcripts have been done in the same way according to the same rules. Sometimes it is sufficient to transcribe only certain parts of an interview, for example if the analysis only focuses on the issues in those parts. Sometimes it is necessary to focus on more detail within a transcript – on how something was said or exactly how long a pause was.

Clarifying what the analysis should reveal

The above decision about the detailing of transcripts should be taken in light of the next step – considering again what the analysis of data should reveal exactly. This also defines here what kind of analysis is necessary. After clarifying the exact question to be answered by the analysis, the various methods for analyzing qualitative data can be applied to transcripts. The data can be analyzed with three kinds of focus: (1) for their contents (what was said?), (2) for formal aspects (how was something said? when was something said?), and (3) for their explicit and implicit meaning. All three kinds of focus can be combined with one or both of the others. We find three kinds of qualitative data analysis: (1) coding allocates statements or observations to categories; (2) language analysis reconstructs how people talk about an issue with one another or what the discourses about an issue are like; and (3) the interpretation of statements or narratives analyzes how experiences are presented in stories.

Qualitative Content Analysis

Against the background of the limitations of quantifying approaches, Mayring (1983) has developed his approach for a qualitative content analysis.

The procedure of qualitative content analysis

The first step here is to define the material (e.g. to select the interviews or those parts that are relevant to answering the research question). You will then analyze the circumstances of data collection (how the material was generated, who was involved, who was present in the interview situation, where the documents to be analyzed come from, etc.). You will continue by formally characterizing the material (whether the material was documented with a recording or a protocol, whether there was an influence on the transcription of the text when it was edited, etc.). Then, you will define the direction of the analysis for the selected texts and 'what one actually wants to interpret out of them' (Mayring 1983, p. 45).

In the next step, the research question is further defined on the basis of theories. A precondition is that the 'research question of the analysis must be clearly defined in advance, must be linked theoretically to earlier research on the issue and generally has to be differentiated in sub-questions' (1983, p. 47). After that, you will select the analytic technique (see below) and define the units. The 'coding unit' defines what is 'the smallest element of material which may be analyzed, the minimal part of the text which may fall under a category'; the 'contextual unit' defines what is the largest element in the text which may fall under a category; the 'analytic unit' defines which passages 'are analyzed one after the other'. Next, you must conduct the actual analyses before you interpret their final results with respect to the research question. Finally, you will ask and answer questions of validity.

Techniques of qualitative content analysis

The concrete methodical procedure essentially involves three techniques of coding. In *summarizing content analysis*, you will paraphrase the material so that you can skip less relevant passages and paraphrases with the same meanings (this is the first reduction) and bundle and summarize similar paraphrases (the second reduction). For example, in an interview with an unemployed teacher, the statement 'and actually, quite the reverse, I was, well, very, very keen on finally teaching for the first time' (Mayring 1983, p. 59) was paraphrased as 'quite the reverse, very keen on practice' and generalized as 'rather looking forward to practice'. The statement 'therefore, I couldn't wait for it, to go to a seminar school, until I could finally teach there for the first time' was paraphrased as 'couldn't wait to finally teach' and generalized as 'looking forward to practice'. Owing to the similarity of the two generalizations, the second one is then skipped and reduced with other statements to 'practice not experienced as shock but as big fun' (1983, p. 59).

Explicative content analysis works in the opposite way. It clarifies diffuse, ambiguous or contradictory passages by involving contextual material in the analysis. Definitions taken from a dictionary or based on grammar are used or formulated. 'Narrow context analysis' picks up additional statements from the text in order to explicate the passages to be analyzed, whereas 'wide context analysis' seeks information outside the text (about the author, the generative situations, from theories, etc.). On this basis, an 'explicating paraphrase' is formulated and tested.

For example, in an interview, a teacher expressed her difficulties in teaching by stating that – unlike successful colleagues – she was no 'entertainer type' (Mayring 1983, p. 109). In order to find out what she wished to express by using this concept, definitions of 'entertainer' were assembled from two dictionaries. The features of a teacher who fits this description were then sought from statements made by the teacher in the interview. Further passages were consulted. Based on the descriptions of such colleagues included in these passages, an 'explicating paraphrase can be formulated: an entertainer type is somebody who plays the part of an extroverted, spirited, sparkling, and self-assured human being' (1983, p. 74). This explication was assessed again by applying it to the direct context in which the concept was used.

With *structuring content analysis*, you look for types of formal structure in the material. You can look for and find four kinds of structure. You may find specific topics or domains which characterize the texts (*content structure*) – for example, xenophobic statements in interviews are always linked to issues of violence and crime. Or you may find an *internal structure* on a formal level which characterizes the material – for example, every text begins with an example and then an explanation of the example that follows. *Scaling structure* means that you find varying degrees of a feature in the material – for example, texts which express xenophobia in a stronger way than other texts in the material. Finally, you may find a *typifying structure* – for example, that interviews with female participants are systematically different from those with male participants in how the main questions are answered. Structuring content analysis consists of four techniques (Mayring 1983, pp. 53–4): formal structuring, which is where formal aspects and an internal structure can be found; content-related structuring, which refers to domains of content; typifying structuring, which looks for salient features of the material; and scaling structuring, which looks for dimensions and scales that can be used to structure the content.

The classification of material in such structures or in dominant categories has to be interpreted, related to the research question and purpose, and assessed for its quality.

Qualitative content analysis can be applied to the analysis of data which comes from interviews, focus groups, newspapers or the Internet.

Box 12.3 Research in the real world

Age and health in journals – an example of a content analysis

In this study (Walter et al. 2006), four journals, two medical and two nursing journals, were content analyzed. The authors selected articles which focused on the topics of health, age, aging, old or very old people, and on prevention and health promotion for the elderly. For this purpose, (1) the content of the title of the articles, (2) the abstract and (3) the whole article were analyzed.

The number of selected articles shows that representations of health, of aging and of old people, as well as the areas of prevention and health promotion for the elderly, are hardly an issue in the four journals. Of the 3028 issues of the journals in the years 1970–2001, only 83 articles explicitly address representations of health; 216 (7.1%) address representations of aging and old people; and 131 (4.3%) mention prevention and health promotion for the elderly, over a period of three decades and in four major journals.

For a more detailed content analysis, 283 publications from 1970 to 2001 were available. The relevance of the issues of aging and of the elderly in the medical and nursing journals was rather low. The distribution of the 216 publications over the period 1970–2001 shows in particular that the general practitioners' journal and the nursing journal pick up this topic, whereas the other two refer to it less often. In the 1970s, only 37 (17.1%) articles could be identified which addressed aging; in the 1980s, 71 (32.9%)

and in the 1990s, 91 (42.1%) referred to the topic. Three of the journals paid more attention to it over time, and only the second nursing journal picked it up in the 1970s more frequently.

In the majority of the 131 identified articles, prevention and health promotion for the elderly are mentioned according to the distinction of primary, secondary and tertiary prevention. Only 12% of the 131 articles address health promotion according to the Ottawa Charter of the World Health Organization (1986).

The quantitative part of prevention and health promotion for older people in most of the articles is below 10%. This means that the topic is mostly only mentioned in one to three sentences. This already shows the marginal relevance of prevention and health promotion in the journals that were studied.

Often, prevention and health promotion are not mentioned explicitly in the selected medical and nursing journals in Germany. This applies in particular to the nursing journals, which use prophylactic aids as a term instead.

All in all, the distribution of the identified publication over the decade shows the following. Representations of health were an issue in the medical and nursing journals in the 1990s less often than in the 1970s. Images of aging and of old people have been increasingly an issue over the decades but without having much relevance. Also, the issues of prevention and health promotion are mentioned more and more. This shows the growing importance of these topics over time, but also that they are still mentioned in a very limited way.

Problems of qualitative content analysis

The schematic elaboration of the procedures makes this method look more transparent, less ambiguous and easier to handle than other qualitative methods of analysis. This is because of the reduction it allows, as outlined above. The method is mainly suitable for reducing large amounts of text by analyzing them on the surface (What is said in them?). Often, however, the application of the rules given by Mayring proves at least as time-consuming as in other procedures. The quick categorization of text based on theories may obscure the view of the contents, rather than facilitate analyzing the text in its depth and underlying meanings. Interpretation of the text (How is something said? What is its meaning?) is applied rather schematically with this method, especially when the technique of explicative content analysis is used. Another problem is the use of paraphrases, which are used not only to explain the basic text but also to replace it – mainly in summarizing content analysis. The more quantitative a content analysis is oriented and applied, the more it reduces the meaning of the text to frequencies and the parallel appearance of certain words or word sequences (see also Schreier 2012 and 2014 for more details on qualitative content analysis).

Interpretations are also relevant in quantitative analysis. Interrelations, which can be found through calculations, such as correlations of two variables, are interpreted by looking for a substantive explanation for these numerical relations. However, the subject of the interpretations is not so much the data themselves but rather the calculations made with them and their results. In the methods that are presented in what follows, interpretation refers immediately to the data and is the actual analysis (see Flick 2014 and 2018a for more details).

Grounded Theory Coding

Whereas qualitative content analysis mainly uses categories developed from the literature and applied to the material to be analyzed, grounded theory coding develops them from the material. Grounded theory coding can be applied to the analysis of data coming from interviews, focus groups, observations, newspapers or the Internet. In the development of a grounded theory from empirical material, coding is the method for analyzing data that have been collected for this purpose. This approach was introduced by Glaser and Strauss (1967) and further elaborated on by Charmaz (2014) and Thornberg and Charmaz (2014). In the process of interpretation, a number of 'procedures' for working with text can be differentiated. They are termed 'open coding', 'axial coding' and 'selective coding'. You should see these procedures neither as clearly distinguishable procedures nor as sequential phases in a linear process. Rather, they are different ways of handling textual material between which the researchers move back and forth, if necessary, and which they combine (see also Chapter 25 in Flick 2018a and 2018e).

The process of interpretation begins with open coding, while, towards the end of the whole analytical process, *selective* coding comes more to the fore. Coding here is understood as representing the operations by which data are broken down, conceptualized and put back together in new ways. Coding includes the constant comparison of phenomena, cases, concepts, and so on, and the formulation of questions that are addressed to the text. *Open coding* aims at expressing data and phenomena in the form of concepts.

For open coding, and indeed for the other coding strategies, it is suggested that you regularly address the text using the following list of so-called basic questions (Strauss and Corbin 1998):

- *What?* What is the issue here? Which phenomenon is mentioned?
- *Who?* Which persons or actors are involved? Which roles do they play? How do they interact?

- *How?* Which aspects of the phenomenon are mentioned (or not mentioned)?
- *When? How long? Where?* Time, course and location.
- *How much? How strong?* Aspects of intensity.
- *Why?* Which reasons are given or can be reconstructed?
- *What for?* With what intention, to which purpose?
- *By which?* Means, tactics and strategies for reaching the goal.

By asking these questions, you may open up the text. They may be applied either to particular passages or to entire texts.

After a number of substantive categories have been identified, the next step is to refine and differentiate the categories that result from open coding. As a second step, Strauss and Corbin suggest completing more formal coding to identify and classify links between substantive categories. In *axial coding*, the relations between categories are elaborated on. In order to formulate such relations, Strauss and Corbin (1998, p. 127) suggest a coding paradigm model, which is symbolized in Figure 12.1.

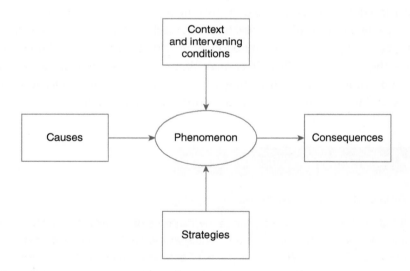

Figure 12.1 The paradigm model

This serves to clarify the relations between a phenomenon, its causes and conse-quences, its context and the strategies of those involved. The coding paradigm outlines possible relations between phenomena and concepts. It is used to facilitate the discovery or establishment of structures of relations between phenomena, between concepts and between categories. Here, as well, the questions asked of the text and the comparative strategies mentioned above are employed once again in a complementary fashion. You will move continuously back and forth between inductive thinking (developing concepts, categories and relations from the text) and deductive thinking. The latter means testing the concepts, categories and relations against the text, especially against passages or cases that are different from those from which they were developed.

In the third step, i.e. *selective coding*, you will focus on elaborating on the potential core concepts or core variables. This leads to an elaboration or formulation of the *story of the case*. In any case, the result should be *one* central category and *one* central phenomenon. You should develop the core category again in its features and dimensions and link it to other categories (all of them, if possible) by using the parts and relations of the coding paradigm. The analysis and development of the theory aim at discovering patterns in the data, as well as in the conditions under which these apply. Grouping the data according to the coding paradigm gives specificity to the theory and will enable you to say, 'Under these conditions (listing them) this happens; whereas under these conditions, this is what occurs' (Strauss and Corbin 1990, p. 131).

Finally, you will formulate the theory in greater detail and again check it against the data. The procedure of interpreting data, like the integration of additional material, ends at the point where *theoretical saturation* has been reached. This means that further coding, enrichment of categories, and so on no longer provide or promise new knowledge. At the same time, the procedure is flexible enough that you can re-enter the same source texts and the same codes from open coding with a different research question and aim at developing and formulating a grounded theory of a different issue.

Box 12.4 Research in the real world

An example of coding – *awareness of dying*

The following represents an important early example of a study that pursued the goal of developing theory from qualitative research in the field. Barney Glaser and Anselm Strauss worked from the 1960s as pioneers of qualitative research and of grounded theory in the context of medical sociology. They did this study in several hospitals in the USA around San Francisco. Their research question asked what influenced various people's interaction with dying people and how the knowledge that the person would die soon determined the interaction with that person. More concretely, they studied which forms of interaction between the dying person and the clinical staff in the hospital, between the staff and the relatives, and between relatives and the dying person could be noted.

The starting point of the research was the observation that, when the researchers' relatives were in the hospital, the staff in hospitals (at that time) seemed not to inform patients with a terminal disease and their relatives about the state or the life expectancy of the patient. Rather, the possibility that the patient might die or die soon was treated as a taboo. This general observation and the questions it raised were taken as a starting point for a more systematic observation and interviews in one hospital. These data were analyzed and used to develop categories. That was also the background for deciding to include another hospital and to continue the data collection and analysis there.

Both hospitals – as cases – were directly compared for similarities and differences. The results of such comparison were used to decide which hospital to use next, until six hospitals were finally included in the study. These included a teaching hospital, a Veterans' Affairs hospital, two county hospitals, a private Catholic hospital and a state hospital. Wards included, among others, geriatrics, cancer, intensive care, pediatrics and neurosurgery, in which the fieldworkers stayed for two to four weeks each. The data from each of these units (different wards in one hospital, similar wards in different hospitals, hospitals amongst each other) were contrasted and compared in order to show similarities and differences.

At the end of the study, comparable situations and contexts outside hospitals and healthcare were included as another dimension of comparison. Analyzing and comparing the data allowed a theoretical model to be developed, which was then transferred to other fields in order to develop it further. The result of this study was a theory of awareness contexts as ways of dealing with the information and with the patients' need to know more about their situation. (Glaser and Strauss 1965)

The method aims at a consistent breaking down of texts. Combining consistently open coding with more and more focused procedures can contribute to a more profound understanding of the contents and meanings of text – one that goes beyond paraphrasing and summarizing (which were the main approaches in qualitative content analysis, discussed above). The advantage is that the interpretation of texts here becomes methodologically realized and manageable. It differs from other methods of interpreting texts because it leaves the level of the pure texts during the interpretation in order to develop categories and relations, and thus theories.

One problem with this approach is that the distinction between method and art becomes hazy. In some places, this makes it difficult to teach or learn as a method. Often, the extent of the advantages and strengths of the method become clear only whilst applying it. If the numbers of codes and possible comparisons become too great, it is suggested that you set up lists of priorities: which codes have to be further elaborated on in all cases, which appear to be less instructive and which can be omitted when you take your research question as a point of reference? (See also Saldana 2013 for more discussion of coding.)

Thematic Coding

If you want to keep the reference to interviewees, for example, as a (particular) case when using a coding procedure, the alternative is to use thematic coding (see Flick 2018a, Chapter 26), which was developed originally to analyze episodic interviews (see Chapter 11). Here, you start your analysis with case studies for which you will develop a thematic structure (what characterizes, across several substantive areas, how the interviewee deals with health? Can you identify issues running through these ways of handling the areas?). In thematic coding, you will first analyze the cases in your study in a number of case studies. For a first orientation, you should develop a short description of each case, which you can continually check and modify throughout the further interpretation of the case, if necessary. This description will include a statement which is typical for the interview, a short characterization of the interviewee with respect to the research question (e.g. age, profession, number of children, if relevant for your issue of research) and the major topics mentioned in the interview with respect to the issue under study. This short description is first a heuristic tool for the following analysis. The example in Box 12.5 comes from my comparative study on everyday knowledge about technological change in different professional groups.

Box 12.5 Research in the real world

An example of a short description of a case

'For me, technology has a reassuring side'

The interviewee is a French female information technology engineer, aged 43 and with a son of 15. She has been working for about 20 years in various research institutes. At present, she works in a large institute of social science research in the computer center and is responsible for developing software, teaching, and consulting employees. Technology has a lot to do with security and clarity for her. To mistrust technology would produce problems for her professional self-awareness. To master technology is important for her self-concept. She narrates a good deal, using juxtapositions of leisure, nature, feeling and family to technology and work, and she repeatedly refers to the cultural benefits of technology, especially television.

In grounded theory coding, you will code material across particular cases in a comparative way from the beginning. In thematic coding, you will go more into the depth of the material by focusing on a single case in the next step (for example, looking at a particular interview as a whole). This single-case analysis has several aims: it preserves the meaningful relations that the respective person deals with in the topic of the study, which is why a case study is done for all interviews; and it develops a system of categories for analysis of the single case.

In the further elaboration of this system of categories (similar to grounded theory coding), first apply open coding and then selective coding. With selective coding, here you will aim less at developing a grounded core category across all cases than at first generating thematic domains and categories for the particular case.

Following the first case analysis, you will cross-check the categories you have developed with the thematic domains that are linked to the single cases. A thematic structure results from this cross-check, which underlies the analysis of further cases in order to increase their comparability.

The structure you developed from the first cases should be continually assessed for all further cases. You should modify it if new or contradictory aspects emerge and use it to analyze all cases that are part of the interpretation. For a fine interpretation of the thematic domains, single passages of text (e.g. narratives of situations) are analyzed in greater detail. The coding paradigm suggested by Strauss (1987, pp. 27–8; see above) is taken as a starting point. The result of this process, complemented by a step of selective coding, is a case-oriented display of the way it specifically deals with the research issue, including oft-repeated topics (e.g. an unfamiliarity with technology) that can be found in viewpoints across different domains (e.g. work, leisure, household).

The thematic structure developed will also serve for comparing cases and groups (that is, for elaborating correspondences and differences between the various groups in the study). Thus, you analyze and assess the social distribution of perspectives on the issue under study.

Thematic Analysis

In the context of psychological research, Braun and Clarke (2006) developed thematic analysis as 'a method for identifying, analysing and reporting patterns (themes) within data. It minimally organizes and describes your data set in (rich) detail. However, frequently it goes further than this, and interprets various aspects of the research topic' (2006, p. 79). It is based on coding and other, more interpretative approaches to texts. The analysis consists of six steps:

1 Familiarizing yourself with your data
2 Generating initial codes
3 Searching for themes
4 Reviewing themes
5 Defining and naming themes
6 Producing the report.

The first step focuses on doing the transcription (yourself) and reading the transcripts several times. In the second step, codes are developed from the material. Here the authors distinguish between semantic codes (meanings expressed verbally) and latent codes (underlying meanings). They suggest working systematically through the whole text, to keep the context of an extract in focus and to keep in mind that statements can be coded in different themes simultaneously. The third step entails sorting the codes into various themes and collating the relevant data extracts in the themes. Step 4 aims at a refinement of the developing codes system by breaking down themes into subthemes, leaving out less relevant ones. In reviewing themes, the researcher can focus either on data extracts or on the entire data set. This should lead, in the fifth step, to thematic maps (visual representations of themes and subthemes and links between them). Then, labels are given to the themes to reflect what they are actually representing. In the sixth (and final) step, the results of this procedure are presented. Thematic analysis can be applied to all sorts of qualitative data, such as interviews, focus groups and Internet data.

Figure 12.2 illustrates how closely analytic methods are oriented on the particularities of the data.

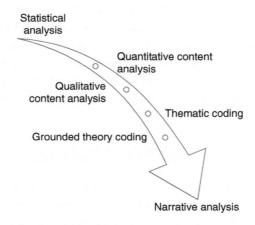

Figure 12.2 Degrees of the method's proximity to the particularities in the data

Interpretative Methods: Analyzing Narratives

In qualitative research, one makes great efforts when collecting the data to be able to understand and analyze statements in their context afterwards. Therefore, in interviews, open questions are asked. In analyzing the data, open coding is used for this purpose, in the first step at least. The analytic methods just discussed, as well as qualitative content analysis, increasingly depart from the original wording of the text: statements assume a new order according to the categories or – in Glaser and Strauss's method – the theories that are developed. More consequently oriented to the Gestalt of the text are methods guided by the principle of sequential analysis.

In the analysis of narrative interviews (see Chapter 11), it is suggested that we first identify all passages of the text where the interviewee gives a narrative presentation and that we focus on them, before including other elements (e.g. descriptions or explanations). Rosenthal and Fischer-Rosenthal (2004) accordingly suggest analyzing narrative interviews in six steps, as follows:

1   Analysis of biographical data (data of events)
2   Text and thematic field analysis (sequential analysis of textual segments from the self-presentation in the interview)
3   Reconstruction of the case history (life as lived)
4   Detailed analysis of individual textual locations
5   Contrasting the life story as narrated with life as lived
6   Formation of types.

Lucius-Hoene and Deppermann (2002, p. 321) have developed a number of questions as a heuristic of analyzing a text:

- What is presented?
- How is it presented?
- For what purpose is this presented – and not something else?
- For what purpose is this presented now – and not at a different time?
- For what purpose is this presented in this way – and not in a different way?

What do the methods for analyzing narrative data discussed above have in common? They take the overall form of the narrative as a starting point for the interpretation of statements, which are seen in the context of the process of the narrative. Furthermore, they include a formal analysis of the material: which passages of the text are narrative passages, which other forms of text can be found?

Problems of narrative analysis

In these methods, the narrative has a varying importance for analysis of the issues under study. If something is presented in the interview in the form of a narrative, Schütze (1983) sees this as an indicator that it has happened in the way it is told. Other authors, however, see narratives as a specifically instructive form of presentation of events and experiences and analyze them as such. Sometimes the assumption is made that narratives, as a form of constructing events, can be met in everyday life and knowledge as well. Thus, this mode of construction can be used for research purposes in a particularly fruitful way. Combining a formal analysis with a sequential procedure in the interpretation of presentations and experiences is typical for narrative analyses in general.

However, particular approaches deriving from Schütze (1983) exaggerate the quality of reality in narratives. The influence of the presentation on what is recounted is underestimated; the possible inference from narrative to factual events in life histories is overestimated. Only in very rare examples are narrative analyses combined with other methodological approaches in order to overcome their limitations. A second problem is how closely analyses stick to individual cases. The time and effort spent on analyzing individual cases restricts studies from going beyond the reconstruction and comparison of a few cases.

Formal Approaches: Talk in (Inter-) Action

The following approaches focus not only on the content of material to analyze but also on the formal aspects – how something is said or how interactions, conversations and discourses are organized and how this influences the construction and transportation of meanings.

Conversation analysis

When applying this method, you will be less interested in interpreting the content of texts than in analyzing the formal procedures with which people communicate and with which specific situations are produced. Classic studies have analyzed the organization of turn-taking in conversations or explained how closings in conversations were initiated by participants. Basic assumptions of conversation analysis are that (a) interaction proceeds in an orderly way and (b) nothing in it should be regarded as random. The context of interaction influences the interaction; the participants of therapy interaction act according to that framework, and the therapist talks like a therapist should do. At the same time, this sort of talking also produces and reproduces this context: by talking like a therapist should do, the therapist contributes to making this situation a therapy and to preventing it turning into a different format of talk – like gossip, for example. The decision about what is relevant in social interaction and thus for the interpretation can only be made through the interpretation and not by *ex ante* settings. Drew (1995, pp. 70–2) has formulated a series of methodological precepts for conversation analysis (CA). He suggests focusing on how talk is organized and in particular on how the speakers organize turn-taking in the conversation. Another focus is on errors and how they are repaired by the speakers. CA looks for patterns of talk and of its organization by comparing several examples of conversations. In presenting the analysis of a conversation, it is important that you give enough examples in verbatim quotes that readers can assess your analysis. For example, if you analyze counseling, you could look at the opening interactions and at how the two participants arrive at defining the issue which the consultation will be about. By comparing several examples, you could show patterns of organizing an issue for the conversation and thus a consultation with a focus.

The procedure of conversation analysis of the material itself involves the following steps. First, you identify a certain statement or series of statements in transcripts as a potential element of order in the respective genre of conversation. The second step is that you assemble a collection of cases in which this element of order can be found. You will then specify how this element is used as a means of producing order in interactions and for which problem in the organization of interactions it is the answer (see Bergmann 2004). This is followed by an analysis of the methods with which those organizational problems are dealt with more generally. Thus, a frequent starting point for conversation analyses is to enquire into how certain conversations are opened and which linguistic practices are applied for ending those conversations in an ordered way (opening up closings).

Research in conversation analysis originally concentrated on everyday conversations (e.g. telephone calls, gossip or family conversations in which there is no specific distribution of roles). Increasingly, however, it has become occupied with specific role distributions and asymmetries as found in, for example, counseling conversation, doctor–patient interactions, and trials (i.e. conversations occurring in specific institutional contexts). The approach has also been extended to include analysis of written texts, mass media or reports, i.e. text in a broader sense (Bergmann 2004; see also Toerien 2014). As Halkier (2010) has shown, this approach can be used to analyze focus groups, including the way the interaction within the groups develops and influences their topics and content of conversation.

Discourse analysis

Discourse analysis has been developed from different backgrounds, one of which was conversation analysis. There are various versions of discourse analysis now available. Discursive psychology, as developed by Edwards and Potter (1992), Harré (1998) and Potter and Wetherell (1998), is interested in showing how, in conversations, 'participants' conversational versions of events (memories, descriptions, formulations) are constructed to do communicative interactive work' (Edwards and Potter 1992, p. 16). There is a special emphasis on the construction of versions of events in reports and presentations. The 'interpretative repertoires' which are used in such constructions are analyzed. Interpretative repertoires are ways of talking about a specific issue. They are called repertoires as it is assumed that these ways are not completely spontaneous, but that people apply certain ways of talking about an issue. At the same time, for example, the way an issue is treated in the press sets up such repertoires (e.g. if a specific ethnic minority is always talked about by referring to violence and crime).

Willig (2003) has described the research process in discourse analysis in several steps. After selecting texts and talk occurring in natural contexts, which have to be described first, you will carefully read the transcripts. Coding and then analyzing the material follows from guide questions like: Why am I reading this passage in this way? What features of the text produce this reading? The analysis focuses on context, variability and constructions in the text and, finally, on the interpretative repertoires used in the texts. The last step, according to Willig, is writing up the discourse analytic research. Writing should be part of the analysis and return the researcher to the empirical material (see also Willig 2014a).

Interpretative Methods of Analyzing Qualitative Data

Common to the interpretative methods discussed above is that they focus on the temporal-logical structure of the text and take this as a starting point in the interpretation (see also Willig 2014b). Thus, they stick more closely to the text than those methods based on categories, which we discussed before. The relation of content and formal aspects is shaped here in different ways. In narrative analyses, the formal difference between narrative and argumentative passages in interviews informs decisions over which passages receive how much interpretative attention and how credible the contents are. Conversation analysis mainly focuses on formal aspects, which are used to design conversations – for example, counseling conversations – and on how they are employed in negotiating the specific contents of a topic. Discourse analysis again takes its turn in analyzing text and talk for content and formal aspects.

Case Studies and Typologies

We now turn to the next step in the model of the analytic process suggested above. This step is focused on the major findings of the analysis and the conclusions they allow, for example by completing a case study or developing a typology.

In Chapter 7, case studies were discussed as one of the basic designs in qualitative research. The methods of data interpretation discussed so far work with case studies in various phases of treating the material. Hermeneutic methods mostly produce a case study in the first stage, consisting of a single interaction, document or interview. Comparing the cases is a later step.

In the approach of grounded theory coding (Glaser and Strauss 1967), however, the single case (the interview, a document or an interaction) is given less attention. When they talk of 'case', they mean instead the research field or issue as a whole. This approach starts immediately by comparing interviewees or specific situations.

Producing and reading case studies

Sometimes, for example in the evaluation of an institution, a case study may be the *result* of the research (see Stake 1995). In other approaches, case studies form the beginning, before comparison becomes more central (see Bassey 1999; Simons 2009). A third possibility is using case studies to illustrate a basically comparative study, in order to highlight links between the different issues studied in the research. In this sense, we included a number of case studies in addition to a thematically structured comparative presentation in our study on the health and illness concepts of homeless adolescents (see Flick and Röhnsch 2008).

An essential point in the production and assessment of case studies is the localization of the case and its analysis. What does the case represent and what do you intend to show by analyzing it? Is the presentation about the single person (or institution, etc.) per se? Does it represent the person as typical for a specific subgroup in the study or does it represent a specific professional perspective (e.g. the physicians in this field)? What were the criteria for selecting this specific case – for the data collection, for the analysis and for the presentation?

Constructing typologies

Typologies are constructed in quantitative research, too. However, this step of condensing and presenting results is more often part of qualitative research. Kelle and Kluge (2010) have made some suggestions for how to apply this, which can be used for developing the following procedure.

The first step is either to construct case studies for the cases that are included in the research or, alternatively, to begin with analyses referring to certain issues. This is followed by systematic comparisons (of cases or referring to issues).

The next step is to define the relevant dimensions for comparison. Consider what the focus of the intended comparison is – for example, the contents of the concepts of health that had been mentioned, on the one hand, and age and gender, on the other. This may reveal here which substantial dimension characterized the ranges of health concepts that were mentioned and how the various dimensions can be allocated in the differentiation of the age of the cases or what can be found more for male or for female interviewees.

The third step is to group the cases (according to the substantial dimension and/or age) and to analyze the empirical regularities (e.g. certain concepts can mainly be found for younger girls, other concepts more for older boys).

As far as possible, the range of statements should be documented and delimited in order to develop a contrastive framework for particular statements. This can be the result of compiling all relevant statements and ordering them along a continuum. In the example of Gerhardt (1988) on family rehabilitation after the husband had fallen chronically ill, this feature space consists of the range of activities found for the husband (professional work; staying at home) and for the wife (professional work; staying at home) and the combinations that result from the four possibilities in the concrete cases. In Gerhardt's study, four types result from this – dual career (both professionally working), traditional (husband working, wife at home), rational (reversed when the husband can no longer work professionally) and unemployment (where no one is working). Gerhardt compared these four types to see in which circumstance rehabilitation of the chronically ill husband was most successful.

The fourth step in the construction of a typology is to analyze substantive meanings. Therefore, you will again analyze the cases in the different types for which meanings of one's own practices can be identified in the interviews and which regularities become visible for this.

Finally, you will characterize the constructed types, exploring which features or combinations of features characterize the cases that have been allocated to the various types: what do they have in common, what distinguishes the cases in the different types? It may be necessary to remove cases strongly deviating from the particular type, to further combine groups in order to reduce the variety or to differentiate into more groups in the case of strong differences within the types constructed so far.

This procedure involves analyzing cases systematically, comparing them and defining a typology. The typology may refer to the cases as a whole, which means to allocate, for example, interviewees to the different types. It may also refer to specific topics, which leads to a typology of how the interviewed adolescents manage sexual risks in their street lives and to a typology referring to their utilization of medical support in the case of health problems. The allocation of the interviewees to these two typologies will not necessarily be identical.

Data Analysis in Mixed Methods and Triangulation Research

Much of what has been discussed for quantitative and for qualitative analysis can be applied in the context of multiple methods approaches. Nevertheless, the linking of mixed methods and in triangulation often happens at the stage of data analysis. Despite the vast set of publications in this field, not much can be found about how to analyze data in mixed methods (e.g. in Creswell 2015).

Independent analysis strategies

Analysis often begins with one approach – the analysis of the quantitative data, for example, is followed by that of the qualitative data, or the other way around. Or both data sets are analyzed in parallel and the results of the two analytic steps are then integrated. These strategies have in common that each set of data is analyzed using the specific form of analyzing data, but that the analysis is not really integrated.

Integrated analysis strategies

Greene (2007, p. 144) has suggested five mixed methods data analysis strategies: (1) data cleaning, which looks for valid responses – methodological soundness, for example; (2) data reduction into descriptive forms such as frequencies, case summaries, descriptive themes, etc.; (3) data transformation, which means the standardization or scaling of quantitative data and the transforming of qualitative data into critical incidents, or chronological narratives (this may also include transforming qualitative into quantitative data and vice versa; see Chapter 9); (4) data correlation and comparison, which look for patterns of relationships in the data set such as clusters of variables, themes or stories that go together across the cases and subgroups of the analysis; in the qualitative part, cross-tabulations of themes, stories, incidents, etc. can serve this purpose; and (5) analyses for inquiry conclusions and inferences, which look for more complex relations in the data – in the quantitative data, for example, by using multivariate analysis of variances and, in the qualitative part, by identifying composite stories, and coherent and cohesive themes. Greene further details these strategies but, ultimately, they consist more of an overarching framework which includes qualitative and quantitative analyses, than a really integrated approach for analyzing mixed methods data. As Bazeley (2010) discusses in more detail, software packages for computer-assisted qualitative analysis such as MaxQDA (see maxqda.com) offer tools for linking qualitative analysis with software packages for statistical analysis such as SPSS.

A suggestion for how to proceed with an integrated analysis strategy

A possible procedure is to begin by identifying the themes and topics, relations and categories in both kinds of data, which are (most) relevant for the study issue and research question. Next, the lists of topics identified in the two data sets in this way are linked and put in parallel order. For each of the topics, the quantitative and qualitative analyses are processed for the data referring to them. Then the results of these processes are linked again by looking for corroborations, for complementing and for contradicting results, before an overall panorama of the findings is created. This will refer to the topics, their links and the data sets.

This suggestion can be used for mixed methods studies using qualitative and quantitative data (e.g. interview and questionnaire data), but it can also be applied to a triangulation of several qualitative data sets (e.g. interview and participant observation data). As outlined in Chapter 9, you should keep in mind that the kinds of data that are combined and analyzed in the above way should refer to different levels of research addressing the overall study topic.

What You Need to Ask Yourself

Box 12.6 What you need to ask yourself

Data analysis

1 Is the method of analysis you chose appropriate to the data you collected?
2 Can you answer your research question with the form of analysis you chose?
3 Did you apply the method of analysis in a consistent way?
4 Are your quantitative results statistically significant?
5 Do any regularities (e.g. a typology) become visible in the qualitative results?
6 How does your analysis consider deviant cases or data?

What You Need to Succeed

If you conduct an empirical project, it will help to consider the following questions for the analysis of your data. These questions can be used in doing your own study or in assessing the research of other researchers.

Box 12.7   What you need to succeed

Data analysis

1   Does the method of analysis you apply fit the method you used for collecting the data?
2   Does it meet the complexity of the data?
3   Does the nature of the data permit you to apply the analytic structure that you chose?
4   Does the kind of analysis allow for a reduction in the complexity of the data in a way that the results become easily understandable?
5   What is the research question or the aspect of it that is intended to be the focus of the analysis?
6   Can your analysis assess whether your result came up by chance or if it is a singular result?

What you have learned

- Content analysis works with texts – in a quantitative way in analyzing newspapers, for example, and in a qualitative way in analyzing interviews and other data.
- Quantitative analysis starts with elaborating on, evaluating and cleaning the data.
- The next step is the descriptive analysis of frequencies, distributions, central tendencies and dispersions in the data.
- Quantitative analyses can focus on one, two or more variables and their relationships.
- Relationships found in this way are tested for their statistical significance with different tests.
- Qualitative analyses can aim at developing a theory by applying various methods of coding the data.
- They can also focus on analyzing narratives for the processes and life histories represented in them.
- The analysis of interactions is an option in analyzing qualitative data.
- Qualitative analysis moves between case-oriented and comparative analysis with the aim of developing typologies.

——What's next——

The following texts provide more detailed discussion of the issues covered in this chapter. In the first resource, quantitative and qualitative analysis are addressed:

Bryman, A. (2016) *Social Research Methods*, 5th edn. Oxford: Oxford University Press.

The following, short book gives a good overview of qualitative data analysis:

Gibbs, G. (2018) *Analyzing Qualitative Data*, 2nd edn. London: Sage.

These next three works focus on how to do qualitative analysis:

Flick, U. (ed.) (2014) *The SAGE Handbook of Qualitative Data Analysis*. London: Sage.
Flick, U. (ed.) (2018) *Doing Grounded Theory*. London: Sage.
Flick, U. (2018) *An Introduction to Qualitative Research*, 6th edn. London: Sage.

In this article, the concept of thematic analysis is spelled out in relation to other methods of data analysis:

Braun, V. and Clarke, V. (2006) 'Using Thematic Analysis in Psychology', *Qualitative Research in Psychology*, 3(2): 77–101.

This next book gives a general orientation to this field:

Rapley, T. (2018) *Doing Conversation, Discourse and Document Analysis*, 2nd edn. London: Sage.

The following case study can be found in the online resources. It should give you an idea of how content analysis can be used to study social media such as Facebook:

Zhao, S. (2014) 'Content Analysis of Facebook Pages: Decoding Expressions Given off', *SAGE Research Methods Cases*. doi: 10.4135/978144627305013511031.

This article, in the online resources as well, emphasizes focusing not only on what was said, but also on how and when it was said:

Halkier, B. (2010) 'Focus Groups as Social Enactments: Integrating Interaction and Content in the Analysis of Focus Group Data', *Qualitative Research*, 10(1): 71–89.

REFLECTION AND WRITING

Successful social research involves much more than merely applying research methods. It is also important to reflect on how methods have been applied – and to make your procedures transparent to others. This final part of the book, therefore, focuses on issues of reflection and writing.

First, we consider quality assessment in quantitative and qualitative research, and then focus on these questions in the context of digital research (Chapter 13).

The final chapter considers how to write up research and results in a transparent way, how to feed them back to participants and how to use data and results in practical or political contexts (Chapter 14).

Research Methodology Navigator

Orientation

- Why social research?
- Worldviews in social research
- Ethical issues in social research
- From research idea to research question

Planning and design

- Reading and reviewing the literature
- Steps in the research process
- Designing social research

Method selection

- Deciding on your methods
- Triangulation and mixed methods

Working with data

- Using existing data
- Collecting new data
- Analyzing data

You are here in your project

Reflection and writing

- What is good research? Evaluating your research project
- Writing up research and using results

13

WHAT IS GOOD RESEARCH? EVALUATING YOUR RESEARCH PROJECT

─How this chapter will help you─

You will:

- be able to identify the most important criteria for evaluating empirical research
- see that these criteria were originally developed for quantitative research
- understand that, for qualitative research, other criteria and approaches of evaluation are applicable
- appreciate the limits of quantitative and qualitative research methods
- know that generalization of results forms a major part in the assessment of social research, and
- be able to distinguish between quantitative and qualitative research with regard to approaches to evaluation.

Evaluating Empirical Studies

As well as the question of the utility of empirical research, we also need to examine its quality. Here, we need to assess whether the methods you applied are reliable, and how far the results you have obtained can claim validity and objectivity. The latter is essentially a question of how far the results could have been obtained by other researchers and are independent of the researcher who did the study. For assessing the quality of empirical research, criteria have been formulated to facilitate assessment of the methodological procedures that led to the results. Here, we can ask whether one uniform set of criteria is adequate for every form of empirical research, or whether we should differentiate between criteria that are appropriate for qualitative or quantitative research. Accordingly, in this chapter we will first discuss the criteria of reliability, validity and objectivity, which are generally accepted in quantitative research, and then consider specific approaches in qualitative research.

A fundamental question here is how far results can be generalized. That is, how far may they be transferred to other situations beyond the research situation? Here again, we should ask whether procedures of or claims for generalization should be formulated in a unified way for all sorts of empirical research, or whether we need differentiated approaches.

Quality and Evaluation of Quantitative Research

In quantitative research, three criteria have been established as core criteria – reliability, validity and objectivity. Its representativeness is a fourth aspect for evaluating a quantitative study.

Reliability

The first generally accepted criterion for assessing studies originates from test theory: 'The reliability ... indicates the degree of exactness in measurement (precision) of an instrument. The reliability is the higher the smaller the part of error E linked to a measurement value X is' (Bortz and Döring 2006, p. 196). We can assess the reliability of a measurement in different ways.

Retest reliability

To assess retest reliability, you will need to apply a measurement (e.g. a test or a questionnaire) *twice* to the same sample and then calculate the correlation between the results of the two applications. In the ideal case, you will obtain identical results. This presupposes, however, that the attribute that was measured is stable in itself and has not changed between the two measurement times. When one is repeating performance tests, differences in the results can arise from a change in performance in the meantime (e.g. because of additional knowledge that was obtained). Differences in the measurements can result because on the second occasion questions are recognized by participants and learning effects may occur.

Parallel test reliability

To assess parallel test reliability, you will need to apply two different instruments, so that you operationalize the same construct in parallel. For example, if you wish to find out how reliable a specific intelligence test is, you can apply a second test in parallel. If the first one has measured the intelligence in a reliable way, the second test should produce the same result – i.e. the same intelligence quotient. 'The more alike the results of the two tests, the less error effects obviously are involved' (Bortz and Döring 2006, p. 197).

Split-half reliability

In a test or a questionnaire, you can calculate the resulting scores for each half of the items (questions) and then compare the two scores. But then the results will depend on the method of splitting the instruments in two halves (e.g. the first and the second half of the questions, even and uneven numbers of questions, random allocation of questions to one or the other half). To exclude effects of the position of the questions, the internal consistency is calculated. To do this, you will treat each question separately like independent tests and compute the correlation between the results, i.e. the answers given to the various questions.

Inter-coder reliability

When you use content analysis, you can calculate inter-coder reliability to assess the extent to which different analysts allocate the same statements to the same categories and hence the reliability of the category system and its application.

Validity

Validity is assessed both for research designs and for measurement instruments.

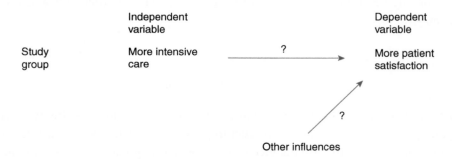

Figure 13.1   Internal validity

Validity of research designs

In the case of research designs, the focus will be on the evaluation of results. You will need to check the *internal* validity of a research design. Internal validity characterizes how far the results of a study can be analyzed unambiguously. If you want to study the effects of an intervention, you should check whether changes in the dependent variables can be traced to changes in the independent variable or whether they may result from changes in some other variable (see Figure 13.1).

Consider, for example, the case of a research project on intensive care, in which the introduction of more intensive care constitutes the independent variable and the satisfaction of the patients the dependent variable. If you wish to study the hypothesis 'more intensive care leads to more satisfaction of the patients', you should clarify how the relation between intensive care and satisfaction can be measured unambiguously. To assess internal validity, you will try to exclude other influences: how far other conditions have changed in parallel to increasing the intensity of care, and how far the increase in patients' satisfaction comes from these conditions.

To ensure internal validity, conditions need to be isolated and controlled. A way to assess the effect of an intervention is to apply a control group design (see Chapter 7). In our example, in a second group, as comparable as possible to the first, the intervention would not be introduced – that is, the intensity of care would not be increased. Then one could check whether the effect found in the study group – i.e. the increase in patients' satisfaction – is evident: 'Internal validity is achieved if the changes in the dependent variables can be unambiguously traced back to the influence of the independent variable, i.e. if there are no better alternative explanations beyond the study hypothesis' (Bortz and Döring 2006, p. 53).

Internal validity is best achieved in the laboratory and in experimental research. However, this is at the expense of the second form of validity of a research design, namely *external* validity. Here, the general question is: how far can we transfer results beyond the situations and persons for which they were produced, to situations and persons outside the research? For example, can we transfer a relation between intensity of care and patients' satisfaction to other wards, hospitals or situations of care in general, or is it only valid within the concrete conditions under which it was studied and found (see Figure 13.2)?

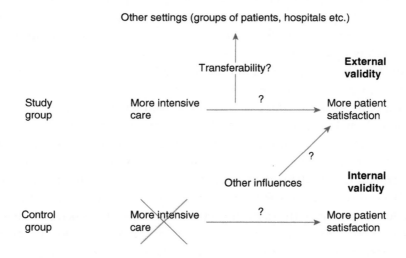

Figure 13.2   External validity

There is a difficulty here. Although in the laboratory and under more or less controlled conditions, internal validity will be high, external validity will in contrast be rather limited. In research in the field and under natural conditions, external validity is higher and internal validity is lower, as here the control of conditions is possible only in a very limited way: 'External validity is achieved when the result found in studying a sample can be generalized to other people, situations or points in time' (Bortz and Döring 2006, p. 53). To meet both criteria in one research design at the same time and to the same extent is seen as difficult (2006). Here, we face a dilemma of empirical research, which is difficult to solve in a research design.

External and internal validity are assessed for research designs. Validity, however, is also assessed for measurement instruments.

Validity of measurement instruments

The issue of the validity of a research instrument can be summarized in the question: does the method measure what it is supposed to measure? To answer this question, you can apply various forms of validity checks, namely (a) content validity, (b) criterion validity and (c) construct validity.

Content validity is achieved when the method or measurement instrument captures the research issue in its essential aspects and in an exhaustive way. You can check this yourself based on your subjective judgment – by reflecting on how far your instrument covers all the important aspects of your issue and whether it does so in a way that is appropriate to the issue. Even better is to have the measurement instrument assessed by experts or laypeople. Errors should catch the eye in such assessments. Thus, the term 'face validity' is used for this assessment. To continue our example: consider whether you would document the intensity of care in the relevant situations of the day routines in a hospital or only in a specific situation – for example, on the admission of patients.

Criterion validity is achieved if the result of a measurement corresponds with an external criterion. This will be the case, for example, if the results of vocational testing correspond with the professional success of the tested person (see Figure 13.3). Such external criteria can be defined in parallel, which allows you to check the concurrent validity. This means that you apply a second measurement at the same time. For example, you do the test and at the same time observe the candidate's behavior in a discussion group. Then you compare the results of the two measurements – how far the part of the vocational test on communicative skills corresponds to the communication in the group. Alternatively, you check the measurement later, in which case the predictive validity will be assessed – for example, do the results of a vocational test allow for the prediction of professional success?

One problem here is that the external criterion has to be valid itself if it is going to be used as a means for checking measurements. Here, you have to take differential validity into account: concordance between the test score and the external criterion can be different in different populations. If we again take our example, the communicative behavior in the discussion groups may be systematically different for male and female participants, whereas the original test mainly focused on general aspects of vocational qualification. The relation between the test score and the external criterion will then be different for the two gender subgroups. Methods in general should be able to capture differences in various groups.

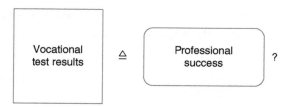

Figure 13.3 Criterion validity

Finally, *construct validity* should be assessed. Here, you will check whether the construct that is captured by your method is linked sufficiently closely to variables that can be theoretically justified. You will also check here how far the construct allows hypotheses to be derived that can be tested empirically. One way to assess construct validity is to use various measurements: constructs are measured using several methods. When several methods measure the same construct with corresponding results, *convergent validity* is given. For example, you may study the patients' satisfaction with a questionnaire and with an interview. When both methods produce results which confirm each other, this shows the convergent validity of your construct. This would indicate that your theoretical concept of 'patients' satisfaction' is valid and that your study meets this criterion of validity. *Discriminant validity* refers to the question of how far your measurements are able to distinguish the construct in your study from other constructs. In our example, you would assess how far your theoretical concept and your measurements really capture the patients' satisfaction with care. Or do they just capture a general state of wellbeing instead of specific satisfaction with aspects of the care situation? In that case, your concept of 'patients' satisfaction' is not valid and your study misses this criterion of validity (see Figure 13.4).

Validity of indices

An index will need to be constructed when something cannot be directly observed or measured (see Chapter 11), because several aspects of a theoretical construct are integrated into it. For example, patients' general satisfaction with their stay in a hospital cannot be measured directly. Such satisfaction includes satisfaction with the treatment, with the staff's friendliness, with the food, with the atmosphere, and so on. To measure the construct 'general satisfaction', you would have to select one or more indicators. In order to reduce bias in the measurement of complex constructs as far as possible, several indicators should be used in order to increase the quality of measurement. (For example, grading a school essay combines evaluations of orthography, style, content and form – either in equal measure or with different weights because style is seen as more important than the number of spelling mistakes.)

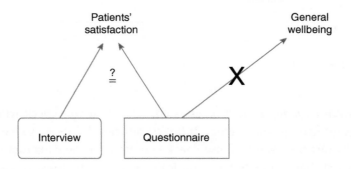

Figure 13.4   Construct validity

In a similar way, one would try to derive patients' satisfaction from various indicators. For example, you could use an instrument to assess quality of life, a questionnaire to measure satisfaction with the service of the hospital staff and another questionnaire to address satisfaction with the infrastructure of the hospital. The question is then how to weigh the particular variables. For example, in constructing a patient satisfaction index, how much weight should be given to the questionnaire results on staff service compared to, say, the results concerning quality of life in the hospital?

A further problem here is that if you want to assess the validity of the index, the included variables – e.g. quality of life, satisfaction with staff – themselves must be measured in a valid way so that the index as a whole can be valid. Thus, validity assessments become relevant on two levels here: on the level of the single indicator, and on the level of the index constructed of these indicators: 'The quality of an index essentially depends on how far all the relevant dimensions have been selected and weighed appropriately' (Bortz and Döring 2006, p. 144). For indices, overall validity is composed of the validity of (a) single items or questions, (b) the scales constructed of these items, and (c) the weighting of the components.

The construction of an index is based on the use of several indicators to measure a value which cannot be directly observed or measured. This raises specific problems concerning the validity of indices. Thus, you should check whether:

- the relevant dimensions have been selected and weighted appropriately
- the instruments for measuring the selected indicators are valid
- the items in the indicators are valid.

Validity addresses different aspects for checking the quality of study results. If you consider external validity, this already includes aspects of the transferability and generalization of results.

Objectivity

The objectivity of instruments, such as tests or questionnaires, depends on the extent to which the application of the instrument is independent of the person applying it. If several researchers apply the same method to the same persons, the results have to be identical. Three forms may be distinguished:

- Objectivity in the data *collection* concerns how far the answers or test results of the participant are independent of the interviewer or person of the researcher. This will be obtained by standardization of the data collection (standardized instructions for applying the instrument and standardized conditions in the situation of the data collection).
- Objectivity of the *analysis* concerns how far the classification of answers in a questionnaire or test is independent of the person who does the classification in the concrete case (e.g. by allocating an answer to a specific score).
- Objectivity of the *interpretation* means that any interpretation of statements or scores in a test should be independent of the person of the researchers and their subjective views or values. Thus, norm values (e.g. age, gender, education) may be identified by representative samples, which can be used to classify the achievement or values of the participant in the concrete study.

Achieving objectivity of an instrument or a study principally requires standardization of the ways in which data are collected, analyzed and interpreted. This will exclude the subjective or individual influences of the researcher or the concrete situation in which data were collected.

Representativeness

This criterion refers to two elements of a study: the sample that was studied and the findings that were obtained. To increase the representativeness of a sample, random sampling is applied. This means that every element of a population has the same chance to be part of the sample in the study, so that bias can be excluded (see Chapter 7). In other forms of sampling, such as quota or purposive sampling (see also Chapter 7), the question has to be asked how far the population is represented in the sample of the study – which means that the results can be generalized from the sample to the population. The representativeness of findings is the second context in which this concept is used, again with an eye to the possibility of generalizing the findings to the population and its elements that are not included in the sample.

Quality and Evaluation of Qualitative Research

The criteria discussed so far are well established for quantitative research. They are based more or less on the standardization of the research situation. It has sometimes been suggested that the classical criteria of empirical social research – reliability, validity and objectivity – may also be applied to qualitative research (see Kirk and Miller 1986). This raises the question of how far these criteria, with their strong emphasis on standardization of procedures and the exclusion of communicative influences by the research, can do justice to qualitative research and its procedures, which are mainly based on communication, interaction and the researcher's subjective interpretations. Often, these bases are seen not as biases but as strengths or even preconditions of the research. Accordingly, Glaser and Strauss 'raise doubts as to the applicability of the canons of quantitative research as criteria for judging the credibility of substantive theory based on qualitative research' (1965, p. 5). They suggest rather that the criteria of judgment be based on generic elements of qualitative methods for collecting, analyzing and presenting data and for the way in which people read qualitative analyses.

In light of such skepticism, a series of attempts has been made over time to initiate a debate about the criteria in qualitative research (see Flick 2018a, Chapter 29). One also finds a number of attempts to develop 'method-appropriate criteria' (see Flick 2018h) in order to replace criteria like validity and reliability.

Reformulation of Traditional Criteria

Suggestions for reformulating the concept of reliability with a more procedural emphasis have focused on the question of how data are produced. A requirement is that (a) statements by participants and (b) interpretation by the researcher should be clearly distinguishable. Finally, one way to increase the reliability of the whole process is to document it in a detailed and reflexive fashion. This refers mainly to documenting and reflecting on the decisions taken in the research process – showing which ones were taken and why (see Chapter 8 for decision-making in the research process).

The concept of validity also requires reformulation. One suggestion is that researchers should scrutinize interview situations for any signs of strategic communication. That means that the interviewee did not openly respond to the questions but was selective or reluctant to give the answers. This leads to a question of how far you can trust the statements of the interviewee. You should also check whether a form of communication occurred in the interview which is not adequate for the interview situation. For example, if the interview produced a therapy-like conversation, this should raise doubts about the validity of the interviewee's statements (Legewie 1987).

A second suggestion is to check validity by integrating participants as individuals or groups into the greater research process. One way is to include communicative validation at a second meeting, after an interview has been conducted and transcribed (for concrete suggestions, see Flick 2018a, p. 229).

Box 13.1   Research in the real world

Using communicative validation

In my study about counsellors' subjective theories of trust (Flick 1992), I applied a semi-standardized interview (see Flick 2018a, Chapter 15) with social workers, psychologists and physicians. I asked them for their subjective definitions of trust in counselling. They were also asked about examples of situations of trustful relations in their work with clients, influences that facilitate or are barriers to developing trust, and the like. Some time (usually about two weeks) after the interview, I met each of the interviewees again at a second meeting. On this occasion, I presented them with the major statements from their first interview. For this, I used little cards with the statements on. Then I asked whether the statements represented correctly what the interviewee had said in the first meeting. This had the purpose of obtaining a communicative validation of the statements. Interviewees could accept, reject or modify these statements. This part of the study was aimed at a communicative validation of the content of the interviews – the data, not the final analysis. But the communicative validation, i.e. the acceptance of the (perhaps modified) statements, is a precondition for further use of the statements for the analysis of the data. The promise of further authenticity made here is twofold. The interviewees' agreement with the content of their statements is obtained after the interview. The interviewees themselves then develop a structure of their own statements in terms of the complex relations the researcher is looking for. This feedback on their statements allowed for the application of communicative validation at the level of content. The visual structuring of the statement allowed the application of this validation at structural and formal levels of knowledge to become visual in the interview.

Sometimes communicative validation is discussed in relation to the validation of the interpretation of texts. Because of the ethical problems in confronting participants with interpretations of their statements, this form of communicative validation is seldom applied. For a more general application of communicative validation, two questions remain to be answered, namely:

- How might one design the methodological procedure in communicative validation (or member checks) in a way that does justice to the issues under study and to the viewpoints of participants?
- How might one answer the question of validity beyond participants' agreement to the data and interpretations?

Mishler (1990) goes one step further in reformulating the concept of validity. He starts from the *process* of validating (rather than from the *state* of validity). He defines 'validation as the social construction of knowledge' (1990, p. 417) by which we 'evaluate the "trustworthiness" of reported observations, interpretations, and generalisations' (1990, p. 419).

Method-appropriate Criteria

The criteria used to assess objectivity need to be appropriate to the methods of qualitative research. Beyond the already mentioned communicative validation and triangulation (see Chapter 9), we find a number of suggestions for new criteria in American discussion (see Flick 2018a, Chapter 29, 2018h). For example, Lincoln and Guba (1985) have proposed (a) trustworthiness, (b) credibility, (c) dependability, (d) transferability and (e) confirmability as the appropriate criteria for qualitative research. Of these, trustworthiness is considered the most important. Lincoln and Guba have outlined five strategies for increasing the credibility of qualitative research:

- Activities for increasing the likelihood that credible results will be produced by a 'prolonged engagement' and 'persistent observation' in the field and the triangulation of different methods, researchers and data
- 'Peer debriefing': regular meetings with other people who are not involved in the research in order to disclose one's own blind spots and to discuss working hypotheses and results with them
- The analysis of negative cases in the sense of analytic induction
- Appropriateness of the terms of reference in interpretations and their assessment
- 'Member checks' in the sense of communicative validation of data and interpretations with members of the fields under study.

To assess dependability, a process of auditing is suggested, based on the procedure of audits in the domain of accounting. The aim is to produce an auditing trail (Guba and Lincoln 1989) covering:

- The raw data, their collection and recording
- Data reduction and results of syntheses by summarizing, theoretical notes, memos and so on, summaries, short descriptions of cases, etc.
- Reconstruction of data and results of syntheses according to the structure of categories (themes, definitions, relationships), findings (interpretations and inferences) and the reports produced with their integration of concepts and links to the existing literature
- Process notes, i.e. methodological notes and decisions concerning the production of trustworthiness and credibility of findings
- Materials concerning intentions and dispositions such as the concepts of research, personal notes and the expectations of participants
- Information about the development of the instruments, including the pilot version and preliminary plans.

In a similar vein, we find a more recent suggestion by Tracy (2010) who proposes eight 'big tent' criteria. She uses this term because the criteria do not refer to a single step in the research process. For example, in a validity check in quantitative research, you will assess the validity of the measurement and put aside other aspects, for instance whether the study addresses a relevant issue at all. Tracy's big tent criteria go beyond such a focus and assess the whole research process. She defines her criteria as follows: 'high quality qualitative methodological research is marked by (a) worthy topic, (b) rich rigor, (c) sincerity, (d) credibility, (e) resonance, (f) significant contribution, (g) ethics, and (h) meaningful coherence' (2010, p. 839). Tracy describes all her criteria in some detail: 'worthy topic', for example, means 'the topic of the research is relevant; timely; significant; interesting'. 'Rich rigor' means 'the study uses sufficient, abundant, appropriate, and complex theoretical constructs; data and time in the field; sample(s); context(s); data collection and analysis processes' (2010, pp. 840–1). Her criterion 'credibility' includes strategies like triangulation and member checks and how deviant cases are treated (here discussed as 'multivocality') (2010, p. 844).

All in all, Tracy makes a suggestion for how to evaluate the approach of the single study as a whole – from the relevance of the research question to the meaningfulness of the results (on theoretical and practical levels) and to research ethics (in the field, in relation to participants and following the end of the project). This distinguishes Tracy's suggestion from the understanding of criteria in quantitative research and proposes a far-reaching and, at the same time, specific concept for evaluating qualitative research.

However, Tracy's suggestions are confronted with the same problem as Lincoln and Guba's suggestions: there are no limits or benchmarks that can be formulated, in terms of how much 'worth', 'rigor', 'credibility' or 'sincerity' should be given in the single study to make it fulfill these criteria. This distinguishes these suggestions of criteria from the 'classical' criteria that allow for definition of such benchmarks in saying a statement is reliable (or not). But this difference is due to the specific characteristics of qualitative research in general and qualitative methods in particular.

The question of transferability of results has been discussed already in the context of external validity. We will address it again below in relation to evaluation.

Generalization

In quantitative research, the extent to which results can be generalized may be checked in two ways – by assessing external validity, one would (a) ensure that the results found for the sample are valid for the population and also (b) test how far they can be transferred to other, comparable populations. Bortz and Döring hold that 'generalizability in quantitative research is achieved by the inference from a random sample (or sample parameters) to populations (or population parameters), which is founded in probability theory' (2006, p. 335).

Various sampling procedures (see Chapter 7) may be used to ensure this. One procedure is to use a random sample, in which every element in the population has the same chance to be an element in the sample. This procedure enables the exclusion of any biases resulting from the disproportionally weighted distribution of features in the sample compared to the population. Thus, the sample is representative for the population. An inference from the sample to the population concerning the validity of the results is therefore justified. Other procedures aim at representing the distribution in the population in a more focused way, as, for example, when you draw a stratified sample. You will then take into account that your population consists of several subgroups, which are unevenly distributed. You will try to cover that distribution in your sample. This allows you to generalize your findings from the sample to the population.

Generalization can be checked by assessing the external validity of a study (see above). This generalization is based on the degree of similarity between participants in the study and the populations for which the study and its results are supposed to be valid. Accordingly, Campbell (1986) uses the term 'proximal similarity' rather than external validity – in the dimensions that are relevant for the study and its results, the sample should be as similar as possible to the population to which the results are to be transferred.

Generalization in qualitative research

In quantitative research, generalization is primarily a numerical problem, to be solved by statistical means. In qualitative research, this question is more difficult. At root there is the familiar issue of generalization: a limited number of cases which have been selected according to specific criteria, or sometimes a particular case, have been studied and the results are claimed to be valid beyond the material in the study. The case or the cases are taken as representative of more general situations, conditions or relations. Yet the question of generalizability in qualitative research often arises in a fundamentally different way. In some qualitative research, the aim is to develop theory from empirical material (according to Glaser and Strauss 1967), in which case the question is raised as to how far the resulting theory may be applied to other contexts.

Accordingly, one approach to evaluating qualitative research is to ask what measures have been taken to define or extend the area of validity of empirical results (and indeed of any theories developed from them). Starting points are the analysis of cases and the inferences from them to more general statements. The problem here is that the starting point is often an analysis focused on a specific context or a concrete case, addressing the specific conditions, relations, processes, etc. It is often precisely the reference to a specific context that gives qualitative research its value. Yet if one then proceeds to generalize, the specific context is lost and one must consider how far the findings are valid independent of the original context.

In highlighting this dilemma, Lincoln and Guba (1985) have suggested that 'the only generalization is: there is no generalization'. Yet in terms of the 'transferability of findings from one context to another' and 'fittingness as to the degree of comparability of different contexts', they outline criteria for judging the generalization of findings beyond a given context. A first step is to clarify the degree of generalization the research is aiming at and which it is possible to attain. A second step involves the cautious integration of different cases and contexts in which the relations under study are empirically analyzed. The generalizability of the results is often closely linked to the way the sampling is done. Theoretical sampling, for example, offers a way of designing a variation of the conditions under which a phenomenon is studied as broadly as possible. The third step consists of the systematic comparison of the collected material.

The constant comparative method

In the process of developing theories, Glaser (1969) suggests the 'constant comparative method' as a procedure for interpreting texts. It consists of four stages: '(1) comparing incidents applicable to each category, (2) integrating categories and their properties, (3) delimiting the theory, and (4) writing the theory' (1969, p. 220). For Glaser, the systematic circularity of this process is an essential feature: 'Although this method is a continuous growth process – each stage after a time transforms itself into the next – previous stages remain in operation throughout the analysis and provide continuous development to the following stage until the analysis is terminated' (1969, p. 220).

This procedure becomes a method of *constant* comparison when interpreters take care that they compare coding over and over again with codes and classifications that have already been made. Material which has already been coded is not finished with after its classification: rather, it is continually integrated into the further process of comparison.

Contrasting cases and ideal type analysis

The process of constant comparison may be systematized further through strategies of contrasting cases. Gerhardt (1988) has made the most consistent suggestions based on the construction of ideal types. This strategy involves several steps. After reconstructing and contrasting the cases with one another, types are constructed. Then 'pure' cases are tracked down. Compared with these ideal types of processes, the understanding of the individual case can be made more systematic. After constructing further types (see Chapter 12), this process culminates in a structural understanding (i.e. the understanding of relationships pointing beyond the individual case).

The main instruments here are (a) the *minimal* comparison of cases that are as similar as possible, and (b) the *maximal* comparison of cases that are as different as possible. They are compared for differences and correspondences.

Generalization in qualitative research involves the gradual transfer of findings from case studies and their context to more general and abstract relations – for example, in the form of a typology. The expressiveness of such patterns can then be specified according to how far different theoretical and methodological perspectives on the issue – if possible by different researchers – have been triangulated and how negative cases were handled. The degree of generalization claimed for a study should also be taken into consideration. Then, the question of whether the intended level of generalization has been reached provides a further criterion for evaluating the qualitative research project in question.

Standards and Quality in Digital Research

All that has been said in this chapter about the criteria and quality of quantitative and qualitative research in principle applies to digital research as well as to traditional research. However, some issues of quality can be raised here with a specific focus on online research. For example, the question of generalization refers not only to how far you can infer from a sample of online users (that were part of a survey) to the population of Internet users in general, but also to how far that specific online sample relates to populations beyond the Internet. There are also specific issues of data protection to take into account. For this reason, extra 'standards for quality assurance for online surveys' have been formulated (see ADM 2001) which are helpful for assessing (your) research with this specific approach.

Limits of Research Approaches

For all the value of social research, it should be recognized that it has its limits. One cannot study everything – for ethical (see Chapter 3), for methodological, and sometimes for practical reasons. Research projects limited to single methods are particularly limited. This chapter is designed both to make those limitations clearer and to show how using a combination of methods can expand the scope of what is possible in research.

Limits of Quantitative Research

Let us first consider the main limitations of quantitative research.

Limits of representativeness

To ensure that studies are representative, researchers need to draw appropriate samples, yet this is easier said than done. In many cases, you will find that biases in the sample arise. For example, you may find that you cannot reach some of the potential participants – they may have moved away or died, for example, or they may refuse to take part in your study. In particular, the increasing number of telephone surveys in the context of market research may result in resistance from potential participants. Sometimes, an increase in funding for a particular area of research results in participants being approached too often and so refusing. As a result, obtaining a representative sample may require considerable effort.

Limits of standardized surveys

Standardized surveys are intended to collect views on the issues under study, independently of both the situation in which data are collected and the person or the interviewer. That is the aim, whether you are using questionnaires to be completed by the participant or standardized interviews with a catalogue of questions to be asked by an interviewer. Yet there are difficulties with these methods. Interviewees' characteristics – for example, their age or gender – can influence the way they approach questions in an interview situation (e.g. Bryman 2016, pp. 216–17). Responses may be influenced by ideas about their social desirability – interviewees may ask themselves, which answer is expected of me? Which opinion would I rather not express? Another problem is that respondents might answer all questions with consistently positive (or consistently negative) answers.

In addition, there is the problem that respondents may interpret questions differently from each other. Can you, for example, assume that everybody who ticks 'agree completely' for a question has understood it in the same way as all the others giving the same answer, and that this answer represents the same attitude in every case? Do interviewee and interviewer share the same meaning of the words in the question? Both would be a precondition for summarizing identical answers under the same category.

Limits of structured observation

For structured observation that uses an observation scheme or guidelines with predefined categories, there are also limitations. Such instruments may force an inappropriate or irrelevant framework upon the setting that you observe. Structured observation documents behaviors rather than underlying intentions and, in general, pays too little attention to their context (Bryman 2016, p. 279).

Limits of quantitative content analysis

Content analyses are only as good as the documents that you study with them. Moreover, it is difficult to develop a category system that does not, to some extent, depend on the interpretation of the coder. Underlying or latent meanings are very difficult to capture through content analysis: such analysis usually remains at a surface level. Also, it is difficult to identify the reasons behind certain statements. Finally, this method has been criticized for its lack of theoretical reference (see Bryman 2016, p. 305 on these points).

Limits in analyzing secondary data

The main limitation of secondary (statistical or routine) data is that researchers may be only partly familiar with the data. You may be over-challenged by the complexity of the data (their volume, their internal structure, etc.). In addition, you will have little control over the quality of the data (how exact was the documentation in the statistics, who conducted the error control or data cleaning and how?). Moreover, sometimes entries or values that are important for your research question may be missing (Bryman 2016, p. 312). In many cases, it will be impossible for you to go back to the raw data, as the data are accessible only in aggregated form (summaries, calculations, distributions). If you use official statistics as data, you will need to ensure that category labels are not misleading and that your definitions are the same as those used in the data collection.

Limits of quantitative research – conclusion

Quantitative research has the advantage that it works with a large number of cases and a clearly structured data basis. Thus, it often provides representative results. The price for this, however, is that data collection and analysis have to be strongly structured and focused in advance. The scope for new aspects, for the specific experiences of individual participants and for taking concrete contexts into account in data collection remains very limited. The data available for the analysis then are significantly reduced in richness due to the way they were collected (e.g. participants may be limited to five alternative responses). Subjective views that fall outside the range anticipated when you designed your instruments will be difficult to handle. This may limit what you are able to study through quantitative methods.

Limits of Qualitative Research

Qualitative research also has its limitations, as described below.

Limits of theoretical sampling

If you use theoretical sampling, you will adapt the selection of material as much as possible to the existing gaps in knowledge. Additional cases are then selected according to which aspects of the research question the analysis has not answered so far. The problem here is that neither the beginning nor the end of selecting material can be defined and planned in advance. Unless you can apply this strategy with much sensitivity, experience and flexibility, the scope of the material that you then analyze, and the generalizability of your results, may remain rather limited.

Limits of interviewing

When you use prepared questions as the core of your interviews, there is a danger that they might omit points that are in fact essential for the interviewees. However, compared to questionnaire studies, you can make this an issue more easily and can repair it more easily. Despite the use of an interview schedule, you will be able to predict what happens in your interviews in a relatively limited way. This also refers to how comparable the situations in your interviews will be. The flexibility in doing the interview permits greater sensitivity to the interviewee, but reduces the comparability of the collected data at the same time. Narratives are even more oriented to particular cases when you collect your data. Therefore, the path from the single interviewee to (theoretically) generalized statements is even longer. The quality of your data will depend very much on your 'success' in the interview situation: do you manage to mediate between what the interviewee mentions and your questions, or to initiate narratives, which include those aspects that are relevant to your study? In any case, the data will be limited to reports about an event or an activity and will not give you direct access to it.

Limits of participant observation

The strength of participant observation is that you, as observer, are really involved in events. Because of your participation, you will have insights into the internal perspective of the setting. As in other forms of observation, participant observation has access only to what happens during the time the researchers participate and observe. What happened before or beyond the setting or concrete situation remains closed to observation and can be covered only through more or less formal conversations. The proximity to what is studied is, at the same time, the strength and the weakness of the method. In many reports (e.g. Sprenger 1989), it is evident that the research situation may overwhelm the researchers; their distance from the situation or people that are observed is then endangered. At the same time, there is the question of how situations, and access to them, are selected. We may ask: how far do the periods in which you make your observations enable you to gain the insights you seek? Do your observations actually elucidate the issue you are researching and show its relevance to the field?

Limits of qualitative content analysis

Compared to quantitative content analysis, there will be, in qualitative research, a stronger consideration of context and the meaning of texts. Compared to other interpretative or hermeneutic approaches, an analysis will be achieved which is based much more on rules and pragmatism. Often, the rules that are formulated for a content analysis are as demanding in their application as other methods. The need to interpret text when classifying may make it difficult to maintain procedural clarity. It puts the promised procedural clarity of the method and its rules in perspective again. Categorization of the material employing categories derived externally or from theories, rather than from the material itself, may direct the researchers' focus more to the content than to exploration of the meanings and depth of the text. Interpretation of text through analytic methods, such as thematic coding or grounded theory analysis, plays a role in qualitative content analysis only in a rather schematic way – in explicative content analysis (as outlined in Chapter 12). Another problem is that, in qualitative content analysis, one often works with paraphrases – these may be useful for explaining the original material but cause problems if used in place of original material for content analysis.

Limits of qualitative analyses of documents

The advantage of documents is that they are often already available, as they have been produced for purposes other than the research. Institutions produce records, notes, statements and other documents that you can use for your research. Again, however, you face the problem that you cannot influence the quality of the data (production). Documents may have other points of focus and different contents than those required for answering your research questions. Sometimes documents have not been produced systematically enough to permit their comparison across institutions. Finally, problems of access to specific documents may arise.

Limits of qualitative research – conclusion

Qualitative research also has important limitations. The very openness, flexibility and richness of qualitative research may make it difficult to make comparisons between data or to see the big picture – 'the wood for the trees'.

Qualitative research may also intrude into the lives and private spheres of participants more than quantitative research does. For example, if a disease has led to a stressful situation in one's life, it may be easier to answer some questions in a questionnaire than to recount in a narrative interview the whole process of becoming ill.

Overall, we can see from the above summary of the typical limitations associated with single research methods that both qualitative and quantitative research, and also particular approaches in both 'camps', have their limits. We now consider whether these limits can be overcome by combining multiple methods

Limits of Triangulation and Mixed Methods in Social Research

The discussion of limits of single methods and of qualitative and quantitative research sought to show that no single method by itself can provide comprehensive access to a phenomenon under study or always be relied upon to provide the appropriate approach. It should also have become clear that quantitative and qualitative approaches in general each have their limitations.

One step towards overcoming the specific limitations of particular methods or approaches is the combination of multiple methods (see Chapter 9). This does not altogether solve the problem of the limitation of research, since each of the methods used remains selective in what it can capture. However, combination or integration can make a research project less restrictive in what it can achieve.

Beyond these limitations, there remain fundamental limits to social research. In addition to the ethical questions discussed in Chapter 3, there are issues that cannot be 'translated' into empirical concepts or approaches. For example, it is not very realistic to try to study the meaning of life empirically. You can, of course, ask participants in a study what for them personally would be the meaning of their lives – but the phenomenon 'meaning of life' in a comprehensive sense will escape an empirical study. Furthermore, social research reaches its limits when immediate solutions are expected. That this is less a methodological than a fundamental problem should have become evident in Chapter 1.

What You Need to Ask Yourself

The following questions should help you to assess your own research or that of other researchers.

> **Box 13.2** What you need to ask yourself
>
> **Evaluating your project**
>
> 1 Is your project consistent in its parts, i.e. does the sampling fit the data collection and both fit the data analysis?
> 2 What did you do with deviant results or negative cases?
> 3 In a quantitative study, how did you assess the representativeness of your sample and of your results?
> 4 In a qualitative study, how did you check the credibility of your research?
> 5 How transparent are your procedures and decision-making within the process?
> 6 Are you aware of any limitations of your methods?

What You Need to Succeed

In the evaluation of an empirical project in social research, the questions in Box 13.3 are designed to inform your practice. These aspects are relevant to evaluating your own research as well as to assessing other researchers' studies.

> **Box 13.3** What you need to succeed
>
> **Evaluating your project**
>
> 1 In a quantitative study, have the criteria of (a) reliability, (b) validity and (c) objectivity been checked?
> 2 In a qualitative study, which criteria or approaches for assessment of quality have been applied?
> 3 How far have you made the study and the assessment of its quality transparent and explicit in your presentation of the results and procedures?
> 4 How have you examined the generalizability of your results? What were the aims in this and how were they reached?
> 5 How did you reflect the limitations of your approach?

> **What you have learned**
>
> - In evaluating quantitative research, the established criteria are reliability, validity and objectivity.

(Continued)

- In qualitative research, these criteria in most cases cannot be applied immediately. Rather, they have first to be reformulated. A number of suggestions have been made for how to reformulate these criteria for qualitative research.
- In addition, method-appropriate criteria have been developed for qualitative research.
- Generalization in quantitative research is based on (statistical) inference from the sample to the population.
- In qualitative research, in contrast, theoretical generalization may be the aim.
- In qualitative research, the building of typologies and the contrasting of cases play a major role.

What's next

The first two texts below discuss quality issues in quantitative research; the other three cover quality in qualitative research:

Bryman, A. (2016) *Social Research Methods*, 5th edn. Oxford: Oxford University Press.
Campbell, D.T. and Russo, M.J. (2001) *Social Measurement*. London: Sage.
May, T. (2001) *Social Research: Issues, Methods and Process*. Maidenhead: Open University Press. Chapter 1.
Flick, U. (2018) *Managing Quality in Qualitative Research*, 2nd edn. London: Sage.
Tracy, S.J. (2010) *'Qualitative Quality*: Eight "Big-Tent" Criteria for Excellent Qualitative Research', *Qualitative Inquiry*, 16: 837–51.

The following case study can be found in the online resources. It should give you an idea of the issues of maintaining data quality that arise in a longitudinal research project:

Smith, C. (2017) 'Longitudinal Research: Transition of Adult Basic Education Students to College', *SAGE Research Methods Cases*. doi: 10.4135/9781526406040.

The following article, in the online resources, demonstrates how the authors attempted to meet one of the more prominent criteria for evaluating qualitative research:

Nowell, L. S., Norris, J. M., White, D. E. and Moules, N. J. (2017) 'Thematic Analysis: Striving to Meet the Trustworthiness Criteria', *International Journal of Qualitative Methods*. https://doi.org/10.1177/1609406917733847.

Research Methodology Navigator

Orientation

- Why social research?
- Worldviews in social research
- Ethical issues in social research
- From research idea to research question

Planning and design

- Reading and reviewing the literature
- Steps in the research process
- Designing social research

Method selection

- Deciding on your methods
- Triangulation and mixed methods

Working with data

- Using existing data
- Collecting new data
- Analyzing data

You are here in your project

Reflection and writing

- What is good research? Evaluating your research project
- Writing up research and using results

WRITING UP RESEARCH AND USING RESULTS

──┤How this chapter will help you├───────────────────────────

You will:

- recognize that the presentation of results and methods is an integral part of social research projects
- know which forms of presentation are appropriate to (a) qualitative and (b) quantitative projects, and
- understand which factors to consider when presenting results to (a) participants and (b) interested institutions and other audiences.

Essentially, social research consists of three steps: (1) planning a study; (2) working with data; and (3) communicating the results. Sometimes writing is even seen as the core of social science:

> Research experiences have to be transformed into texts and to be understood on the basis of texts. A research process has findings only when and as far as these can be found in a report, no matter whether and which experiences were made by those who were involved in the research. (Wolff 1987, p. 333)

When you communicate your research findings, you should aim to make the process that led you to them transparent to the reader. In the process, you should aim to demonstrate that your findings are not arbitrary, singular or questionable, but rather that they are based on evidence. This chapter considers the relevance of research, the utilization of its results and their elaboration and presentation.

Goals of Writing Up Social Research

When you report your research, you may have various aims. For example:

- To document your results
- To show how you proceeded during your research – how you arrived at your results
- To present results in such a way that you can achieve specific objectives – for example, to obtain a qualification, to influence a process (such as through evidence-based policy-making) or to show something fundamentally new in your field of study
- To legitimize the research in some way – by showing that the results are not arbitrary but rigorously based on data.

Writing Up Quantitative Research

In quantitative research, a research report will normally include the following elements:

- The research problem
- The conceptual framework
- The research question and the hypothesis to be tested
- The method of data collection
- The analysis of the data
- Conclusions
- Discussion of the results.

When presenting your research, you should include statements referring to each of these points. They should permit an assessment of your methodology, the robustness of your results and their relationship to previous literature and research.

The empirical procedure

When writing about your empirical procedure, you should answer the following questions for your readers (see Neuman 2014, p. 521–22):

1   What type of study (e.g. experiment, survey) have you conducted?
2   How exactly did you collect the data (e.g. study design, type of survey, time and location of data collection, experimental design used)?
3   How were variables measured? Are the measures reliable and valid?
4   What is your sample? How many subjects or respondents are involved in your study? How did you select them?
5   How did you deal with ethical issues and specific concerns of the design?

The essential point is to allow the reader to evaluate your study. This will make it possible to assess the relevance and reliability of your results.

Figure 14.1 illustrates the three areas that writing about research should address: (1) the planning and methodology of a study; (2) the fieldwork done for it; and (3) the results and presentation of the study as such. For each area, elements are listed that should be covered.

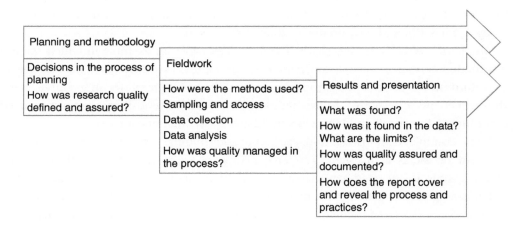

Figure 14.1   The process of writing about research

Elaboration of results

Often, researchers face the problem that a study may produce a multitude of findings. Here, the first step is to provide a number of detailed findings. You can then distill, extrapolate or construct the fundamental results that emerge from the detailed findings. For example, you may do more to sort the data into classes.

Even the researchers themselves may feel confused by the multitude of findings produced in their study. In this case, they need to develop a framework for continuing analysis. For example, they may focus on the distribution of data or they may adopt a comparative perspective (e.g. focusing on the differences between two subgroups in the sample). Above all, the researchers will need to be selective. As Neuman explains:

> You must make choices in how to present the data. When analyzing the data, you look at dozens of univariate, bivariate and multivariate tables and statistics to get a feel for the data. This does not mean that you include every statistic or table in a final report. Instead, select the minimum number of charts or tables that fully inform the reader. Use data analysis techniques to summarize the data and test hypotheses (e.g. frequency distributions, tables with means and standard deviations, correlations, and other statistics). (2014, p. 521)

Presenting results

Frequencies and numbers can often be presented more clearly in charts than in words. Consider, for example, Figure 14.2, which presents the sample sizes in a longitudinal study at three times of measurement (2006, 2007, 2008).

An alternative is to use pie charts. Figure 14.3 provides an example in which the age distribution of homeless people in Germany is summarized in age groups (e.g. 30–39 years).

A third means of presentation is to use tables. Table 14.1 provides an example. Here, the frequencies of responses to the question of how to prevent disease are summarized according to the age of children, who could give more than one answer.

These examples demonstrate how you can present findings visually so that they become apparent for readers 'at first sight'. The three methods above are not, of course, the only ones available; they are included here merely to illustrate the usefulness of visual presentation.

For example, Rew et al. (2007) analyzed the outcomes of a short intervention concerning the sexual health of homeless adolescents in a control group design (see Chapter 7), with repeated measures before (control condition) and during the intervention: 'Two hundred and eighty-seven adolescents participated only in the control condition phase of the study, 196 participated only in the intervention phase of the study and 89 participated in both the control and the intervention phases of the study' (2007, p. 820). The authors used the data in Table 14.1 to demonstrate their results concerning the sample composition in the intervention and comparison (control) group, and those who participated both in the comparison and intervention phases.

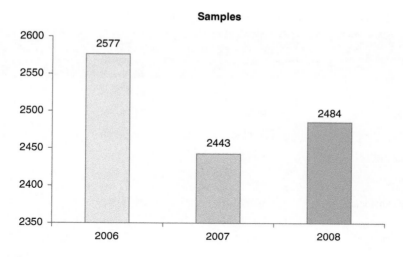

Figure 14.2 Bar chart

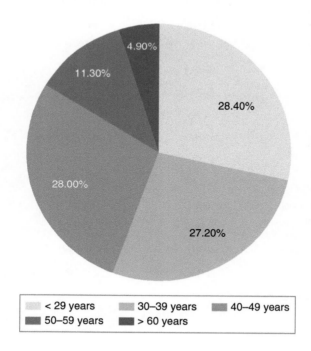

Figure 14.3 Pie chart

Table 14.1 Demographic characteristics of homeless adolescents in comparison and intervention groups

	Comparison conditions (n = 287)	Intervention (n = 196)	Both (n = 89)
Sex			
Female	94 (33%)	78 (40%)	33 (37%)
Male	181 (63%)	113 (58%)	52 (58%)
Did Not Answer	12 (4%)	5 (2%)	4 (5%)
Race/Ethnicity			
African American/ Black	21 (7%)	21 (11%)	7 (8%)
Asian American	4 (1%)	2 (1%)	0 (0%)
European American/ White	168 (59%)	112 (57%)	49 (55%)
Latino/Hispanic	34 (12%)	18 (9%)	14 (16%)
American Indian	21 (7%)	8 (4%)	2 (2%)
Multiethnic/Multiracial	22 (8%)	17 (9%)	8 (9%)
Other	9 (3%)	12 (6%)	5 (6%)
Did Not Answer	8 (3%)	6 (3%)	4 (4%)
Sexual Orientation			
Bisexual	63 (22%)	35 (18%)	19 (21%)
Gay	4 (1%)	2 (1%)	2 (2%)
Lesbian	3 (1%)	5 (3%)	0 (0%)
Straight/Heterosexual	197 (69%)	136 (69%)	55 (62%)
Transgender	3 (1%)	0 (0%)	3 (3%)
Uncertain/Questioning	11 (4%)	11 (6%)	7 (8%)
Did Not Answer	6 (2%)	7 (4%)	3 (3%)

Source: Rew et al. 2007, p. 821

Evidence

In many areas of research, the notion of evidence-based practice has become very important. This emphasizes the need to distinguish reliable from less reliable scientific results.

Here we can consider the example of medical practice. In evidence-based medicine, a medication is tested in a specific form of research before it is introduced into regular medical routines. Greenhalgh gives the following as a definition:

> Evidence-based medicine is the use of mathematical estimates of the risk of benefit and harm, derived from high-quality research on population samples, to inform clinical decision making in the diagnosis, investigation or management of individual patients. The defining feature of evidence-based medicine, then, is the use of figures derived from research on *populations* to inform decisions about *individuals*. (2006, p. 1)

Here, double-blind tests (see Chapter 7) are applied, which are also labeled randomized controlled studies (RCTs). Participants in a drug study are allocated randomly to the treatment group (with the medication) and a control group (with a placebo) to test the effect of the medication by comparing the two groups. On the existence of evidence in the area of the epidemiology of disease (e.g. lung cancer) in relation to certain risk factors (e.g. smoking), Stark and Guggenmoos-Holzmann hold: 'If certain criteria of scientific evidence are fulfilled, a causal relation between the influential variable (risk factor) and the target variable (disease) is generally accepted' (2003, p. 417). They list a number of major criteria, such as 'A strong correlation ... between the influential and the target variables ... results can be confirmed in several studies (reproducibility); a logical temporal course of cause and effect, plausible results and: If the risk factor is eliminated, the risk of illness is reduced' (2003, p. 417).

Thus, a very specific understanding of what evidence is and, moreover, what kind of research can produce such evidence, has been developed. More generally, evidence-based practice means that professional practice and decision-making in the single case should be based on research and results, i.e. on evidence. This scientific foundation of professional practice should replace decision-making based on anecdote ('I knew a similar case ...') or on what is fashionable in the press.

Here, we may note two implications of the development just mentioned. First, this understanding of evidence and research threatens to push back other approaches of research and to challenge their relevance. Consequently, we find several suggestions for classifying types of evidence. Such classification ranges from meta-analyses, based on randomized studies, as the most accepted form, to case studies and expert evaluations, which are the forms with the least acceptance (see Chapter 7).

Second, other disciplines – like social work, education, nursing and others – have also been developing evidence-based practice in their field. The trend towards evidence basing threatens to question other forms of research. There is a problem here: what makes a lot of sense in assessing the effect of medications is not necessarily justified for social research in general. Other forms to be replaced by evidence are concepts of cost minimizing, etc. (Greenhalgh

2006, pp. 9–11). The challenge is to develop a broader understanding of what evidence is. Thus, analyzing the life histories of cancer patients, for example, will not meet the criteria of evidence basing in the sense outlined above, but will produce helpful evidence when it comes to understanding how people live with cancer and how they try to cope with it.

Writing Up Qualitative Research

For qualitative research, many commentators have expressed doubts that we can use the same approach as for quantitative research. Lofland (1976) and Neuman (2014, p. 523) have proposed an alternative structure for reporting qualitative research:

1   Introduction:

    (a) most general aspects of the situation
    (b) main contours of the general situation
    (c) how materials were collected
    (d) details about the setting
    (e) how the report is organized

2   The situation:
    (f) analytical categories
    (g) contrast between the situation and other situations
    (h) development of the situation over time

3   Strategies of interaction
4   Summary and implications.

This structure provides a possible framework for your report. The advantage is that you will then go from more general aspects of your research issue to the more concrete procedures and, finally, to the findings you obtained. Thus, this structure contributes to leading your readers through your report and directing them to the central points you want to make about your findings.

The empirical procedure

For qualitative projects, the presentation of empirical procedures is often chronological, providing in the process insights into the field or situation of study. Van Maanen (1988) distinguishes three forms of presentation, namely (a) 'realist', (b) 'confessional' and (c) 'impressionist' tales.

In *realist tales*, one reports observations as facts, or documents them by using quotations from statements or interviews. Emphasis is laid on the *typical* forms of what is studied (see Flick et al. 2010 for an example). Therefore, you will analyze and present many details. Viewpoints of the members of a field or of interviewees are emphasized in the presentation: how did they experience their own life in its course? What is health for the interviewees? The interpretation does not stop at subjective viewpoints but goes beyond them with various and far-reaching interpretations.

Confessional tales are characterized by the personalized authorship and authority of the researcher as an expert. Here, the authors express the role that they played in what was observed, in their interpretations and also in the formulations that are used. The authors' viewpoints are treated as an issue in the presentation, alongside the problems, breakdowns, mistakes, etc. (Van Maanen 1988, p. 79) in the field. Nevertheless, authors will attempt here to present their findings as *grounded* in the issue that they studied. Such reports combine descriptions of the studied object and of the experiences of studying it. An example of this kind of report is Frank's book *The Wounded Storyteller* (1997).

Impressionist tales take the form of dramatic recall. The aim is to place the audience imaginatively in the research situation, including the specific characteristics of the field and of the data collection. A good example of this kind of report is Geertz's (1973) analysis of the Balinese cockfight.

Results

Qualitative studies can produce various forms of results. They may range from detailed case studies to typologies (e.g. several types of health concepts) or to the frequency and distribution of statements in a category system. In qualitative research, one may again be able to condense information into the form of tables (see Box 14.1).

Box 14.1   Research in the real world

Using tables for presenting qualitative research

In our study on homelessness and health (see Flick and Röhnsch 2007), we used tables to communicate the research and its results, as the following examples show. We gave an overview of the composition of the sample regarding gender and age groups in years (see Table 14.2).

(Continued)

Table 14.2 Sample street youth by age and gender

Age (in years)	Adolescents		
	Male	Female	Total
	N = 12	N = 12	N = 24
14 – 17	5	9	14
18 – 20	7	3	10
Mean	17.5	16.0	16.75

Source: Flick and Röhnsch 2007, p. 739

A second example clarifies the concrete meaning of being homeless here by illustrating what the adolescents said about where they were currently staying overnight (see Table 14.3).

Table 14.3 Current housing situation

Currently staying overnight	Adolescents		
	Male	Female	Total
	N = 12	N = 12	N = 24
At a friend's or an acquaintance's place	9	6	15
In assisted living	3	4	7
At a relative's or partner's place	–	2	2

Source: Flick and Röhnsch 2007, p. 740.

In qualitative research, this has mostly the aim of contextualizing the particular statements and their interpretations (see Table 14.4, which summarizes the health definitions given by the adolescents).

Such presentations in table form are less a substitute for presenting the research in the detail of results or for interpretations of what was said by whom and of what the meaning behind that is, and more a provision of the wider context for seeing the statements and interpretations and their limits.

Table 14.4 Health concepts of the street youth interviewees

Concept of health	Adolescents		
	Male	Female	Total
	N = 12	N = 12	N = 24
Health as physical and mental wellbeing	6	3	9
Health as a result of practices	2	5	7
Health as absence of illness	1	3	4
Health as functionality	3	1	4

Source: Flick and Röhnsch 2007, p. 740

This form of presentation takes a kind of comparative overview approach – what was said in a context ('What is health for me?') and how to classify it (types of health concepts in the variety of answers and to whom or which subgroups in the sample these statements or types apply) (see Table 14.4). To maintain the overall context of cases in the study, this was complemented by short case studies. These case studies summarize for the particular participant major statements in the areas of the interview and the links among them. Such case studies are about 6–10 pages long. Therefore, we included such case studies only for a few selected cases of particular significance for the field or the results.

One specific form of result (or outcome) of a qualitative study may be the development of a theory. The presentation of such a theory requires, according to Strauss and Corbin (1990, p. 229), four components:

1   A clear analytic story.
2   Writing on a conceptual level, with description kept secondary.
3   The clear specification of relationships among categories, with levels of conceptualization also kept clear.
4   The specification of variations and their relevant conditions, consequences, and so forth, including the broader ones.

Here, presentation of research will highlight the core concepts and lines of the theory that have been developed. Visualization in the form of conceptual networks, trajectories and so on is a means of giving the presentation more pregnancy. The suggestions of Lofland (1974) for presenting findings in the form of theories lead us in a similar direction. He mentions as criteria for writing the same criteria as used in evaluating such reports, namely ensuring that:

(1) The report was organized by means of a generic conceptual framework; (2) the generic framework employed was novel; (3) the framework was elaborated or developed in and through the report; (4) the framework was eventful in the sense of being abundantly documented with qualitative data; (5) the framework was interpenetrated with the empirical materials. (1974, p. 102)

In qualitative research, one again faces the problem of selecting essentials from a multitude of data and relationships that the study has identified. Here, you must find a balance in your report between providing insight into, on the one hand, the occurrences and conditions in the field and, on the other, the data and more general issues you can derive from such insights and details. This is necessary for making your results accessible and transparent to readers. One way is to combine topic-related summaries and structures with case studies, which go beyond those topics and show commonalities between the topics. A major issue in presenting qualitative research is how to make the data trail from raw data to categories and presentation transparent to readers. In an interview study, you cannot present every statement (not even every interesting statement) in your article or thesis. You cannot present every step your coding process went through, including all the initial and preliminary codes you developed, later abandoned or integrated with other codes. And, finally, you will have to select the findings that are most relevant to your study (or, e.g., the part of it you currently want to present in a paper). The art of writing about this kind of research is how to find a way to allow readers to develop their own picture and understanding of your work – not only how to convince them but also how to allow them to assess what you present (and found). For this purpose, it will be helpful to present excerpts and examples demonstrating the coding and selection process and to include elements of earlier steps in the analysis. In grounded theory research, it might be helpful to present (a limited number of) memos or to describe the career of codes and categories from open to selective coding (and into the presentation of the study).

Evidence in qualitative research

The concept of evidence according to the evidence-based approaches outlined above cannot be transferred to qualitative research without problems. However, the question has to be asked of how to define evidence in the use of qualitative methods, and what role qualitative research can play in this development (see Morse et al. 2001; Denzin and Lincoln 2005). The discussion includes the suggestion of linking several studies – in the sense of meta-analyses – and also extending and transferring results to practical contexts as a form of 'test' (Morse 2001). For example, we found in one of our studies (Flick et al. 2010) three types of knowledge about the link between residents' sleeping problems at night and the lack of daytime activities in the nursing home for nurses. One of them was rather limited; one of them was much more developed. We can use these results to train nurses and then assess whether this intervention changes their practices and reduces the intensity of sleeping problems. In this way, practice provides a kind of 'test' concerning the evidence of our findings.

In the context of Cochrane Intervention reviews on the effectiveness of intervention in the healthcare context, four possibilities are discussed for how to make use of qualitative research (Hannes 2011, p. 2):

- Used to define and refine review questions in the Cochrane Review (informing reviews);
- Use identified whilst looking for evidence of effectiveness (enhancing reviews);
- Use of findings derived from a specific search for qualitative evidence that addresses questions related to an effectiveness review (extending reviews);
- Conducting a qualitative evidence synthesis to address questions other than effectiveness (supplementing reviews).

This will become more relevant, the more qualitative research is also subject to meta-analysis (Timulak 2014) and reanalysis (Wästerfors et al. 2014). Here, the synthesis of qualitative evidence will become more relevant as an approach to integrating and presenting findings:

> Qualitative evidence synthesis is a process of combining evidence from individual qualitative studies to create new understanding by comparing and analysing concepts and findings from different sources of evidence with a focus on the same topic of interest. Therefore, qualitative evidence synthesis can be considered a complete study in itself, comparable to any meta-analysis within a systematic review on effects of interventions or diagnostic tests. (Higgins and Green 2011, n.p.)

New forms of presenting qualitative research

In several contexts, discussion has started on how to present qualitative research beyond the writing up of data in the 'classical' formats of articles and books, referring to other people and situations and how they were analyzed and with what results. First, we find an autoethnography trend: 'Autoethnography is an approach to research and writing that seeks to describe and systematically analyze (*graphy*) personal experience (*auto*) in order to understand cultural experience (*ethno*)' (Ellis et al. 2011, para 1). This trend is not only reframing formats of presentation but also ways of doing research. At the same time, functions of writing are seen as therapeutic for readers and participants (2011, para 26). Second, we find a trend in moving on from writing about research to forms of performative social science (see the special issue in the journal *Forum Qualitative Social Science* [2009/2] and Guiney Yallop et al. 2009). Here again, it is less the form of presentation of findings, which has been reframed but rather the understanding of qualitative research in general (e.g. the roles of methods and data in the process).

Issues of Writing

There are some general issues concerning the writing of research, regardless of the type of research project. In producing a text, researchers need to consider their potential readers or audience:

> Making your work clearer involves considerations of audience: who is it supposed to be clearer to? Who will read what you write? What do they have to know so that they will not misread, or find what you say obscure or unintelligible? You will write one way for the people you work with closely on a joint project, another way for professional colleagues in other specialities and disciplines, and differently yet for the 'intelligent layman'. (Becker 1986, p. 18)

The choice of audience will have implications for the style of presentation. One needs to ask, what can you assume as already existing knowledge and available terminologies on the part of the readers? How complex and detailed should the account be? Which forms of simplification are specifically necessary for my audience?

A major issue in writing about research is how far the research allows answering the research question of the project and how far the report about it provides the information necessary for judging whether it has been answered and how.

If you are writing about your research in the context of a thesis or an assignment, it is most important that you refer to the department's or faculty's rules of writing, formatting and presenting, and adapt your presentation to these rules as early as possible to avoid problems and unnecessary extra work.

Outlets for Writing

There are several outlets you can use for writing about your research. The classical formats are journal articles and books. Publishing in books can mean writing a monograph about your study (e.g. Glaser and Strauss 1965), its issue, procedures and findings or about a methodological approach (e.g. Glaser and Strauss 1967) that was developed in this context. It can also mean editing a book with contributions by other authors complementing your own chapters in it, or contributing a chapter to a book edited by other researchers. In most social sciences, we can observe a shift from the book as a format of writing about research to the journal article. This has to do with the current preference for making journal contents available on databases like Web of Science and the Social Science Citation Index (SSCI). A second reason is that journal articles are in most cases subject to quality assessments before publication, in particular peer-review processes. And, finally, the evaluation of scientific outputs in general is more focused on journals, articles and their citations than on books and book chapters. In the field of journals, we can distinguish between traditional journals published by a commercial publisher in printed form and sold individually or in the form of subscriptions (to libraries or individual subscribers) and electronic journals. Now, most printed journals are also available electronically – you can buy single articles or subscribe to the journal's electronic edition.

For several reasons, the electronic form of journal publishing has been complemented by a different type of journal, which is known as 'open access'. This means that these journals can be accessed free of charge by everyone. FQS – *Forum Qualitative Social Science* (www.qualitative-research.net/) – is an example of such an open-access journal. Note that there are differences between journals in the field of open-access publishing – such journals may or may not have a serious peer-review process; they may be free for readers but charge contributors for the peer-review process and/or for publishing an article. These charges can be quite expensive in some cases. They may be more or less accepted in the scientific community. For example, it can be instructive to find out who the editors are and who is on the editorial board, or who has published in the journal you are interested in. Some journals now offer both – traditional publishing for which the readers pay and open-access publishing for which the contributors pay.

Beyond books and journals, there are other formats you can use to disseminate your findings. For example, you could decide to print a number of copies of your report or thesis and distribute them at your own cost to relevant target groups (personal publishing). Or you could start publishing your results at conferences as oral papers or posters. Bude summarizes the general problem:

> One is made aware that scientific knowledge is always presented as scientific knowledge. And the consequence is that a 'logic of presentation' has to be considered as well as a 'logic of research'. How researchers' constitution of experiences is linked to the way those experiences are saved in presentations has only begun to become an issue for reflection and research. (1989, p. 527)

Feeding Back Results to Participants

In practice-oriented research, feedback of preliminary or final results to the participants is often expected – particularly if the results are likely to be used for evaluation or decision-making. In this case, you will need to consider how to present your research transparently in a form that participants will find accessible. Furthermore, you should take into account the dynamics in the field – what the results may initiate or change in this context. When describing how she fed back the results and evaluations in her studies in the context of police research, Mensching (2006) used the term 'mediating work': her research had to be presented very sensitively in order to avoid being either too challenging or too simplistic. You must also ensure that you protect particular participants from being identified from the results – even by their colleagues.

Box 14.2 Research in the real world

Feedback of results to participants

In our study on health professionals' social representations about health and aging (Flick et al. 2004), we organized the feedback of our results to our participants as follows. Following episodic interviews with particular participants about their health concepts and ideas and their experiences with prevention and health promotion, and after analyzing the data, we ran focus groups with general practitioners and home-care nurses. The results of the study were reported back to participants to collect their comments on the findings. We then discussed practical implications with them concerning improvements in the routines and practices of home-care nursing and medicine.

In order to avoid discussion in groups becoming too general or heterogeneous, we selected a concrete part of the data as a stimulus for opening up more comprehensive discussions. For this purpose, we selected results concerning the barriers to a stronger orientation of one's own practice towards prevention. Results were presented on the readiness for and resistance to more prevention on the part of both professionals and patients. First, we gave an overview of the barriers that had been mentioned. Then participants were asked to rank these barriers according to their relevance to them, before discussing this relevance for their own professional practice and for the role of health within it. When this discussion abated, we asked the participants to make suggestions about how to overcome the barriers discussed before, and again to discuss these suggestions. At the end, we had not only the evaluations of the results of the initial interviews but also a list of comments and suggestions from every group. These lists could be analyzed and integrated into the final results of the study.

Using Data in Debate

The specific problem of using data in debates has been highlighted by Lüders (2006). There are questions concerning how far the data and the conclusions drawn from them are credible and can be trusted:

> From the viewpoint of the political administration … the question simply is: can we also rely on the data and the results in the classical understanding of reliability and at least internal validity? … The questions of credibility and of trust refer not merely to the problem of whether the data are reliable and valid for the units that were studied, but mainly to the problem of their transferability to similar contexts. (2006, p. 456)

This issue is heightened when the data come from a qualitative study with very few cases and where political initiatives may be based on the results. Here, the issue is the robustness of the results in the process of argumentation and the implications that follow:

> When in a small qualitative study with seven cases in two areas of a country the result is that juvenile intensive offenders – and in these seven cases the concept is adequate – keep the responsible organizations (police, legal services, child and youth services, mental services for children and adolescents, schools, etc.) so busy that no one has an overview anymore, then this result is first valid for these seven cases without doubt. The *politically* relevant question, however, is: are the results robust enough that a process is inaugurated using them to redefine the institutional responsibilities in the country for such a constellation – and how is this defined? – with all the necessary effort, extending even to changing the laws? This also refers to the risks linked to those changes, for example of conflict between interest groups that are concerned and the possible advantage handed to the political opposition in parliament. (Lüders 2006, pp. 458–59)

Lillis (2002, p. 36) has argued in a different context (i.e. market research) that the movement from data to practical suggestions or implications entails three steps:

- The first step focuses on what participants (in qualitative studies) think and mean. By analogy, you can see what the numbers in a quantitative study say (e.g. what correlates with what). This refers to the report on the research and the immediate results;
- The second step in both cases relates to what patterns emerge and what they mean. This is mainly based on the interpretation of the data and results by the researchers;

- The third step asks for the implications for the commissioner of the study and thus for the applicability of the results for them.

For research-based consultation, we should distinguish between those suggestions which (1) are grounded in the research alone, (2) involve knowledge about the relevant market beyond the results, and (3) additionally take the commissioners' specific product or service into account (2002, p. 38). Suggestions will only rarely be based on the empirical results alone; if they are to have any effect, they will have to take into account the current political situation and the specific mandate of the institution.

Utilization of results

It should be recognized that the results of a research project might not necessarily be taken up one-to-one in the different contexts in which they are read. Rather, they will be subject to further interpretation. Studies about the utilization of social science research results (e.g. Beck and Bonß 1989) have shown that various processes of using, reinterpreting, evaluating and selecting occur in utilization:

> Since the results of the social science utilization research, we know that the utilization of scientific knowledge more or less occurs according to the logic of the respective field of practice – in our case the logic of political administration in a way that is often rather confusing for the scientists. (Lüders 2006, p. 453)

As we have seen in Chapter 13, the evaluation of the research is no longer conducted using only the internal criteria of sciences alone. Rather, it becomes evident that the 'quality criteria and checks of the research ... are no longer exclusively defined out of the disciplines and with "peer reviews", but that additional or even competing social, political and economic criteria emerge from the contexts of application (of the results)' (Weingart 2001, p. 15).

If social science and its results are intended to have practical, political or other forms of impact, such processes must be taken into account. The results and the scientific quality of research alone will not prove decisive. Rather, results have to be elaborated on and presented in such a way that they become relevant and understandable for the specific context of discussion and application. For qualitative research, as well as for mixed methods, it becomes a challenge: how to prove the relevance of qualitative research findings. An important issue in this context is how research findings are presented, so that they reach and convince readers outside the qualitative research community. Sandelowski and

Leeman (2012) made that point quite clear when they published their article 'Writing Usable Qualitative Health Research Findings'. They present a number of suggestions for how to make research presentations accessible, for example: 'The key strategy for enhancing the accessibility and usability of qualitative health research findings is to write in the language of the readers to whom they are directed' (2012, p, 1407). Murray (2014) has discussed in some detail ways of implementing results of (qualitative) research in practical contexts.

What You Need to Ask Yourself

The questions in Box 14.3 should help you to reflect on your presentation of the study you did.

Box 14.3 What you need to ask yourself

Presenting research

1 Have you made transparent, how the idea of this research came about?
2 Have the contexts of the project been made clear?
3 Can readers understand why you selected the methods you applied?
4 Is it transparent, how you applied them and if so how you adapted them?
5 Have you discussed the limits of your research, of your methods and of the results?

What You Need to Succeed

When presenting an empirical project in social research, you should take the aspects listed in Box 14.4 into account. These questions again are relevant not only in writing your own report but also in assessing how other researchers present their research.

Box 14.4 What you need to succeed

Presenting research

1 Have the aims of the project been made clear?
2 Has the research question been explicitly formulated and grounded?

(Continued)

---

3 Is it evident why particular persons or situations have been involved in the study and what the methodological approach of the sampling is?

4 Is it evident how the data were collected, for example by presenting example questions?

5 Is it transparent how data collection proceeded and which special occurrences played a role – perhaps for the quality of the data?

6 Will readers be able to understand how the data were analyzed?

7 Have questions (criteria and strategies) of quality assurance in the research been addressed?

8 Have the results been condensed so that their essentials become evident to readers?

9 Have consequences been drawn from the study and the results, and discussed?

10 Is the report easy to read and understand and has the text been complemented by illustrations?

11 Did you provide enough evidence (e.g. quotations or extracts from the calculations) to enable the readers to evaluate your results?

What you have learned

- Social research and its results become accessible only through the form of writing that you choose.
- The aim is to make transparent to readers both how the researchers proceeded and what results they obtained.
- There is a need to select and weigh what you found in order to direct your readers' attention to what is essential in the results.
- Quantitative research and qualitative research differ in the way they present their findings and in the processes that have led to them. However, the two types of research also have some aims in common.
- Particular needs of elaborating on data and results pertain to political or administrative consultations and the feedback presented to participants.
- The utilization of results often follows a logic other than those of research and researchers.

What's next

The works listed below provide further discussion on the issues of writing and presenting social research and evidence. This first book addresses the challenges and techniques of writing about research:

Becker, H.S. (1986) *Writing for Social Scientists*. Chicago: University of Chicago Press.

This next textbook has in Chapter 28 a detailed and helpful part on writing up quantitative research:

Bryman, A. (2016) *Social Research Methods*, 5th edn. Oxford: Oxford University Press.

The following book has in Chapter 30 a detailed discussion on writing up qualitative research:

Flick, U. (2018) *An Introduction to Qualitative Research*, 6th edn. London: Sage.

This next text focuses on reading empirical research:

Greenhalgh, T. (2006) *How to Read a Paper: The Basics of Evidence-Based Medicine*. Oxford: Blackwell-Wiley.

The following two case studies can be found in the online resources. The first should give you an idea about using systematic reviews for informing a practical and political audience about research:

Virdun, C., Luckett, T., Lorenz, K., Davidson, P. and Phillips, J. (2017) 'Analyzing Consumer Priorities for Hospital End-of-Life Care Using a Systematic Review to Inform Policy and Practice', *SAGE Research Methods Cases*. doi: 10.4135/9781526424884.

This second case study should give you an impression of the chances and limits of representing a specific (sub-) culture in the findings of research and their presentation:

LaVoulle, C. (2015) 'Above the Drum: A Study of Visual Imagery used to Represent the Changes in Hip-hop', *SAGE Research Methods Cases*. doi: 10.4135/9781446273050145361 25.

The following two articles, in the online resources, address the problems of presenting qualitative research based on quotations and using case studies for displaying complex issues and studies:

Taylor, S. (2012) '"One Participant Said ...": The Implications of Quotations from Biographical Talk', *Qualitative Research*, 12(4): 388–401.
Seidelsohn, K., Flick, U. and Hirseland, A. (2020) 'Refugees' Labor Market Integration in the Context of a Polarized Public Discourse', *Qualitative Inquiry*, 26(2): 216–226.

GLOSSARY

Applied research Different from basic research, studies are done in specific practical fields and theories are tested for contexts of use in practice (e.g. the hospital).

Assessment (In medical or psychological contexts) measurement of the (cognitive) performance or the health and illness status of patients, for example.

Audit Strategy to assess a process (in accounting or in research) in all its steps and components.

Background theories Theories that inform social research approaches through their conceptualization of reality or research.

Basic research Research that starts from a specific theory in order to test its validity without restriction to a specific field of application or topic.

Big Data This term refers to the huge amounts of data that result from digital communication, administration and the linking of various social networks.

Bivariate analysis Calculations showing the relation between two variables.

Burnout A syndrome of exhaustion by one's own profession, often caused by high stress and a lack of positive feedback.

Categorization Allocation of certain events to a category. Summarizing several identical or similar events under a concept.

Central tendency The average in a distribution (e.g. mean, median).

Closed question A finite range of answers are presented, often in the form of a scale of answer possibilities.

Code of ethics A set of rules of good practice in research (or interventions) established by professional associations or by institutions as an orientation for their members.

Coding Development of concepts in the context of grounded theory – for example, labeling pieces of data and allocating other pieces of data to them (and the label). In quantitative research, coding means allocating a number to an answer.

Coding paradigm A set of basic relations for linking categories and phenomena among each other in grounded theory research.

Communicative validation A criterion for validity for which the consent of the study participants is obtained for data that were collected, for interpretations or for results.

Complete collection A form of sampling that includes all elements of a population defined in advance.

Constant comparative method Part of grounded theory methodology focusing on comparing all elements in the data with each other. For example, statements from an interview about a specific issue are compared with all the statements about this issue in other interviews and also with what was said about other issues in the same and other interviews.

Constructionism A variety of epistemologies in which a social reality is seen as the result of constructive processes (activities or the members or processes in their minds). For example, living with an illness can be influenced by the way that individuals see their illness, what meaning they ascribe to it and how this illness is seen by other members of their social world. On each of these levels, illness and living with it are socially constructed.

Context sensitivity Methods that take the context of a statement or observation into account.

Contingency analysis Analysis of how often certain concepts appear together with other concepts in the press during the study period.

Control group design A design that includes two groups, of which one (the control group) will not receive the treatment being studied in order to control whether the effects that are observed in the treatment group also occur without treatment or whether they really can be located in the treatment.

Conversation analysis Study of language (use) for formal aspects (e.g. how a conversation starts or ends or how turns from one to the other speaker are organized).

Correlation General concept for describing relations between variables, e.g. number of divorces and amount of income.

Covert observation A form of observation in which the observers do not inform the

field and its members of the fact that they are doing observations for research purposes. This can be criticized from an ethical point of view.

Critical rationalism Epistemology which justifies the test of hypotheses by falsification.

Data protection Conduct or means that guarantee the anonymity of research participants and ensure that data do not end up in or are passed into the hands of unauthorized persons or institutions.

Deduction A logical reference from the general to the particular or, in other words, from theory to that which can be observed empirically.

Dependent variable A variable that belongs to the 'then' part of a (if–then) hypothesis and which shows the effects of the independent variable (cause, effects).

Description Where studies provide (only) an exact representation of a relation or of the facts and circumstances.

Dispersion A measure for the variation in measurements.

Document analysis Where existing documents (files, protocols, images, websites) are used as data and analyzed.

Double-blind test An empirical study in which participants and researchers do not know the aim of the study. A 'blind' researcher does not know whether she is in contact with a member of the study group or the control group. This avoids the participants being unconsciously influenced.

Episode A short situation in which something specific happens, such as a period of illness.

Episodic interview A specific form of interview which combines question–answer sequences with narratives (of episodes).

Episodic knowledge Knowledge based on memories of situations and their concrete circumstances.

Epistemology Theories of knowledge and perception in science.

Ethics committees Committees in universities, or sometimes also in professional associations, that assess research proposals (for dissertations or funding) for their ethical soundness. If necessary, these committees pursue violations of ethical standards.

Ethnography A research strategy combining different methods but based on participation, observation and writing about a field under study. For example, to study how homeless adolescents deal with health issues, a participant observation in their community may be combined with interviews of the adolescents. The overall image of details from this participation, observation and interviewing is unfolded in a written text about the field. The way of writing gives the representation of the field a specific form.

Ethnomethodology A theoretical approach interested in analyzing the methods people use in their everyday life to make communication and routines work.

Evaluation research The use of research methods with a focus on assessing a treatment or intervention for demonstrating the success or reasons for failure of the treatment or intervention.

Everyday theory Theories that are developed and used in everyday practice for finding explanations and making predictions.

Evidence-based practice An intervention (in medicine, social work, nursing, etc.) that is based on results of research done according to specific standards.

Experiment An empirical study in which certain conditions are produced deliberately and observed in their effects. Participants are distributed randomly to experimental and control groups.

Expert interview A form of interview that is defined by the specific target group – people in certain professional positions, which enables them to provide information about professional processes, or a specific group of patients, for example.

Explanation Identification of regularities and relations in experience and behavior by assuming external causes and testing their effects.

External validity A criterion of validity focusing on the transferability of results to other situations beyond the research situation.

External variable An influence other than that to be studied.

Factor analysis A form of statistical analysis, focusing on identifying a limited number of basic factors and summarizing them so that they can explain the relations in a field.

Falsification Disproof of a hypothesis.

Field notes Notes taken by the researcher on their thoughts and observations when they are in the field 'environment' they are researching.

Field research Studies done not in a laboratory but in practical fields – like a hospital – in order to analyze the phenomena under study in real conditions.

Focus group A research method used in market and other forms of research, in which a group is invited to discuss the issue of a study for research purposes.

Generalizability The degree to which the results derived from a sample can be transferred to the population.

Generalization Transfer of research results to situations and populations that were not part of the research situation.

Grounded theory Theories developed from analyzing empirical material or from studying a field or process.

Hermeneutics The study of interpretations of texts in the humanities. Hermeneutical interpretation seeks to arrive at valid interpretations of the meaning of a text. There is an emphasis on the multiplicity of meanings in a text, and on the interpreter's foreknowledge of the subject matter of a text.

Heuristic Something that is used for reaching a certain goal.

Homogeneous sample A form of sampling in which all elements of the sample have the same features (e.g. age and profession).

Hypothesis An assumption that is formulated for study purposes (mostly coming from the literature or an existing theory) in order to test it empirically in the course of the study; often formulated as an 'if–then' statement.

Ideal type A pure case that can represent the phenomenon under study in a specifically typical way.

Independent variable A variable that belongs to the 'if' part of a (if–then) hypothesis.

Index construction A combination of several indicators, which are collected and analyzed together (e.g. the social state is composed of the states of education, profession and income).

Indication A decision about when exactly (under what conditions) a specific method (or combination of methods) should be used.

Indicator Something representing a specific phenomenon that is not directly accessible.

Induction Reference from the specific to the general or, in other words, from empirical observation to theory.

Informed consent Agreement by the participants in a study to be involved in it that is based on information about the research.

Internal validity A form of validity that defines how unambiguously a relation that was measured can be captured, i.e. how far biases coming from external variables can be excluded.

Intervening variable A third variable that influences the relation between independent and dependent variables.

Interview A systematic form of asking people questions for research purposes – either in an open form with an interview schedule or in a standardized form similar to a questionnaire.

In vivo code A form of coding based on concepts taken from an interviewee's statements.

Key person A person that makes access to the field under study (or to certain persons in it) possible or easier for the researcher.

Likert scale A five-step scale for measuring attitudes, for which certain statements are presented and agreement with those statements is collected.

Longitudinal study A design in which the researchers return repeatedly, after some time, to the field and the participants to do interviews several times again in order to analyze developments and changes.

Market research Use of research methods in analyzing the market for specific products.

Measurement Allocation of a number to a certain object or event depending on their degree, according to defined rules.

Member check Assessment of results (or of data) by asking participants for their consensus.

Mixed methodology An approach combining qualitative and quantitative methods in a pragmatic manner.

Multivariate analysis Calculation of the relations between more than two variables.

Narrative A story told by a sequence of words, actions or images, and more generally the organization of the information within that story.

Narrative analysis A study of narrative data which takes the context of the whole narrative into account.

Narrative interview A specific form of interview based on one extensive narrative. Instead of asking questions, the interviewer asks participants to tell the story of their lives (or of, e.g., their illness) as a whole, without interrupting them with questions.

Naturally occurring data Data that are not produced by special methods (like interviews) but are only recorded in the way they can be found in the everyday life under study.

Objective hermeneutics A way of doing research by analyzing texts to identify latent structures of meaning underlying those texts and explain the phenomena that are the issues of the text and the research. For example, analyzing the transcript of a family interaction can lead to identifying and elaborating on an implicit conflict underlying the communication of the members in this interaction and on other occasions. This conflict as a latent structure of meaning shapes the members' interaction without their being aware of it.

Objectivity A criterion for assessing whether a research situation (the application of methods and their outcome) is independent of the single researcher.

Open question A form of question in a questionnaire for which no pre-formulated responses are provided and which can be answered using keywords or a short paragraph written by the respondent.

Operationalization Means for empirically covering the degrees of a feature, for which you will define the methods of data collection and measurement.

Paradigm A fundamental conception of how to do research in a specific field with consequences as to the levels of methodology and theory.

Parallel test reliability A criterion of reliability for an instrument which is tested by applying a second measurement instrument in parallel.

Paraphrase A reformulation of the core of the information included in a specific sentence or statement.

Participant observation A specific form of research in which the researcher becomes a member of the field under study in order to make observations.

Participative research A form of research in which the study participants are integrated into the design of the study. The aim is change through research.

Participatory action research Research that intends to produce changes in the field under study. This is to be achieved by planning interviews in a specific way and by feeding back results to participants. Members of the field are made active participants in the design of the research process.

Patients' career The stages of a series of treatments of a chronic illness in several institutions.

Peer debriefing A criterion of validity for which colleagues' comments about the results of a study are obtained.

Peer review Assessment (of papers before publication) by colleagues from the same discipline acting as reviewers.

Population Aggregate of all possible study objects about which a statement is intended (e.g. nurses). Research studies mostly include selections (samples) from a population and results are then generalized from samples to populations.

Positivism A philosophy of science which bases the latter on the observation of data. The observation of data should be separated from the interpretation of their meanings. Truth is to be found by following general rules of method, largely independent of the content and context of the investigation.

Pragmatic test of theories A form of theory assessment in everyday life, when people test their implicit theories by looking at how far they can be used to explain certain relations.

Pre-test Application of a methodological instrument (questionnaire or category system) with the aim of testing it before its use in the main study.

Principle of openness A principle in qualitative research according to which researchers will mostly refrain from formulating hypotheses and instead formulate (interview) questions as openly as possible in order to come as close as possible to the views of participants.

Protocol A detailed process documentation of an observation or a group discussion. In the first case, it is based on the researchers' field notes; in the second case, interactions in the group are recorded and transcribed, often complemented by researchers' notes on the features of communication in the group.

Qualitative method A research method aiming at a detailed description of processes and views that are therefore used with small numbers of cases in the data collection.

Quantitative method A research method aiming at covering the phenomena under study in their frequencies and distribution and thus working with large numbers in the data collection.

Questionnaire A defined list of questions presented to every participant in a study in an identical way either written or orally. The participants are asked to respond to these questions mostly by choosing from a limited number of alternative answers.

Random sample A sample drawn according to a random principle from a complete list of all elements in a population (e.g. every third entry on a page of the telephone directory), so that every element in the population has the same chance to be integrated into the study. This is the opposite of purposive or theoretical sampling.

Reactivity Influences on persons in a study due to their knowledge about being studied.

Reconstruction The identification of continuous topics or basic conflicts in, for example, a patient's life history.

Reliability A standard criterion in quantitative research which is based on repeated application of a test to assess

whether the results are the same in both cases.

Representative sample A sample that corresponds to the major features of the population – for example, a representative sample of all Germans corresponds in its features (e.g. age structure) to the features of the German population.

Representativeness A concept referring to the generalization of research and results. It is understood either in a statistical way (e.g. is the population represented in the sample in the distribution of features such as age, gender, employment?) or in a theoretical way (e.g. are the study and its results covering the theoretically relevant aspects of the issue?).

Research design A systematic plan for a research project, specifying whom to integrate into the research (sampling), whom or what to compare for which dimensions, etc.

Research method Everyday techniques such as asking, observing, understanding, etc., but used in a systematic way in order to collect and analyze data.

Response rate The number or share of questionnaires that are completed and returned in a study.

Sample A selection of study participants from a population according to specific rules.

Sampling A selection of cases or materials for study from a larger population or variety of possibilities.

Scaling The allocation of numerical values to an object or event.

Secondary analysis The use of existing datasets that were produced for other purposes.

Secondary data Data that have not been produced for the current study but are available from other studies or from, for example, the documentation of administrative routines.

Segmenting Decomposition of a text into the smallest meaningful elements.

Semantic knowledge Knowledge organized around concepts, their meaning and in relation to each other.

Semi-structured interview A set of questions formulated in advance, which can be asked in a variable sequence and perhaps slightly reformulated in the interview in order to allow interviewees to unfold their views on certain issues.

Sequential analysis Analysis of a text that proceeds from beginning to end, along the line of development in the text, rather than by categorizing it.

Significant relation A relation that (statistically) is stronger than the estimation of its non-existence.

Standard deviation A root drawn from the variance.

Standardization Control of a research situation by defining and delimiting as many features of it as necessary or possible.

Study design The plan according to which a study will be conducted, specifying the selection of participants, the kind and aims of the planned comparisons and the quality criteria for evaluating the results.

Survey A representative poll.

Systematic review A way of reviewing the literature in a field that follows specific rules, is replicable and transparent in its approach and provides a comprehensive overview of the research in the field.

Thematic coding An approach involving analysis of data in a comparative way for certain topics after case studies (e.g. of interviews) have been done.

Theoretical sampling The sampling procedure in grounded theory research, where cases, groups or materials are sampled according to their relevance to the theory that is developed, and against the background of what is already the state of knowledge after a certain number of cases have been collected and analyzed.

Theoretical saturation The point in grounded theory research at which more data about a theoretical category do not produce any further theoretical insights.

Theory development The use of empirical observations to derive a new theory.

Theory test Assessment of the validity of an existing theory on the basis of empirical observations.

Transcription The transformation of recorded materials (conversations, interviews, visual materials, etc.) into text for analysis.

Triangulation The combination of different methods, theories, data and/or researchers in the study of one issue.

Typology A form of systematization of empirical observation by differentiating several types of a phenomenon, allowing the bundling of single observations.

Utilization research Studies analyzing how research results are adopted in practical contexts.

Validity A standard criterion in quantitative research, for which you will check, for example, whether confounding influences affected the relations under study (internal validity) or how far the results are transferable to instances beyond the current research situation (external validity).

Variable How cases differ or vary in terms of a specific attribute.

Variance Sum of the square deviation of the values from the mean.

Verstehen The German word for 'to understand'. It describes an approach to understanding a phenomenon more comprehensively than reducing it to one explanation (e.g. a cause–effect relation). For example, to understand how people live with a chronic illness, a detailed description of their everyday life may be necessary, rather than merely identifying a specific variable (e.g. social support) to explain the degree of success in their coping behavior.

Virtual ethnography Ethnography through the Internet – for example, participation in a blog or discussion group.

Vulnerable population People in a specific situation (e.g. social discrimination, risk, illness) who require particular sensitivity when they are being studied.

Working hypothesis An assumption that is made to orient the work in progress. Different from hypotheses in general, this is not tested in a standardized way.

REFERENCES

Abdulai, R. T. and Owusu-Ansah, A. (2014) 'Essential Ingredients of a Good Research Proposal for Undergraduate and Postgraduate Students in the Social Sciences', *SAGE Open*. https://doi.org/10.1177/2158244014548178

ADM (2001) *Standards for Quality Assurance for Online Surveys*. Available at: www.adm-ev.de/index.php?id=2&L=1 (accessed 2 July 2010).

Albert, H. (1992) 'Kritischer Rationalismus', in H. Seiffert and G. Radnitzky (eds) *Handlexikon zur Wissenschaftstheorie*. München: Oldenbourg. pp. 177–82.

Alheit, P., Rheinländer, K. and Watermann, R. (2008) 'Zwischen Bildungsaufstieg und Karriere Studienperspektiven "nicht-traditioneller Studierender"', *Zeitschrift für Erziehungswissenschaft*, 11 (4): 577–606.

Allmark, P. (2002) 'The Ethics of Research with Children', *Nurse Researcher*, 10: 7–19.

Allport, G. (1954) *The Nature of Prejudice*. New York, NY: Basic Books.

Anderson, P. (2007) *What is Web 2.0? Ideas, Technologies and Implications for Education*. JISC Technology and Standards Watch, February. Bristol: JISC. Available at: www.jisc.ac.uk/media/documents/techwatch/tsw0701b.pdf (accessed 29 October 2010).

Andrews, R. (2003) *Research Questions*. London: Continuum.

Asmus, J. (2017) *Diskussionsbeteiligung in Geschichtsseminaren – eine Analyse der Beteiligung mit dem Fokus Geschlecht*. Unpublished master's thesis, Freie Universität Berlin.

Baker, S.E. and Edwards, R. (2012) 'How Many Qualitative Interviews is Enough?', Discussion Paper, National Center of Research Methods, http://eprints.ncrm.ac.uk/2273/

Bampton, R. and Cowton, C.J. (2002) 'The E-Interview', *Forum: Qualitative Social Research*, 3 (2). Available at: www.qualitative-research.net/fqs/fqs-eng.htm (accessed 22 February 2005).

Banks, M. (2018) *Using Visual Data in Qualitative Research*, 2nd edn. London: Sage.

Barbour, R. (2014) 'Analyzing Focus Groups', in U. Flick (ed.) *The SAGE Handbook of Qualitative Data Analysis*. London: Sage. pp. 313–26.

Barbour, R. (2018) *Doing Focus Groups*, 2nd edn. London: Sage.

Barton, A.H. and Lazarsfeld, P.F. (1955) 'Some Functions of Qualitative Analysis in Social Research', *Frankfurter Beiträge zur Soziologie*. Frankfurt: Europäische Verlagsanstalt. pp. 321–61.

Bassey, M. (1999) *Case Study Research in Educational Settings*. Milton Keynes: Open University Press.

Baur, N. and Florian, M. (2009) 'Stichprobenprobleme bei Online-Umfragen', in N. Jackob, H. Schoen and T. Zerback (eds) *Sozialforschung im Internet: Methodologie und Praxis der Online-Befragung*. Wiesbaden: VS-Verlag. pp. 109–28.

Baym, N.K. (1995) 'The Emergence of Community in Computer-Mediated Communication', in S. Jones (ed.) *Cybersociety: Computer-Mediated Communication and Community*. London: Sage. pp. 138–63.

Bazeley, P. (2010) 'Computer Assisted Integration of Mixed Methods Data Sources and Analyses', in A. Tashakkori and C. Teddlie (eds) *Handbook of Mixed Methods Research for the Social and Behavioral Sciences*, 2nd edn. Thousand Oaks, CA: Sage. pp.431–68.

Beck, U. and Bonß, W. (eds) (1989) *Weder Sozialtechnologie noch Aufklärung? Analysen zur Verwendung sozialwissenschaftlichen Wissens*. Frankfurt: Suhrkamp.

Becker, H.S. (1986) *Writing for Social Scientists*. Chicago: University of Chicago Press.

Becker, H.S. (1996) 'The Epistemology of Qualitative Research', in R. Jessor, A. Colby and R.A. Shweder (eds) *Ethnography and Human Development*. Chicago: University of Chicago Press. pp. 53–72.

Berger, P.L. and Luckmann, T. (1966) *The Social Construction of Reality*. Garden City, NY: Doubleday.

Bergmann, J. (2004) 'Conversation Analysis', in U. Flick, E. v. Kardorff and I. Steinke (eds) *A Companion to Qualitative Research*. London: Sage. pp. 296–302.

Bergmann, J. and Meier, C. (2004) 'Electronic Process Data and their Analysis', in U. Flick, E. v. Kardorff and I. Steinke (eds) *A Companion to Qualitative Research*. London: Sage. pp. 243–7.

Bishop, L. and Kuula-Luumi, A. (2017) 'Revisiting Qualitative Data Reuse: A Decade On', *SAGE Open*. https://doi.org/10.1177/2158244016685136

Black, A. (2006) 'Fraud in Medical Research: A Frightening, All-Too-Common Trend on the Rise', *Natural News*. Available at: www.naturalnews.com/019353.html (accessed 1 September 2010).

Blaikie, N. and Priest, J. (2019) *Designing Social Research*, 3rd edn. Cambridge: Polity Press.

Blumer, H. (1969) *Symbolic Interactionism: Perspective and Method*. Berkeley and Los Angeles: University of California Press.

Bogner, A. and Menz, W. (2009) 'The Theory-Generating Expert Interview', in A. Bogner, B. Littig and W. Menz (eds) *Interviewing Experts*. Basingstoke: Palgrave Macmillan. pp. 43–80.

Bogner, A., Littig, B. and Menz, W. (eds) (2009) *Interviewing Experts*. Basingstoke: Palgrave Macmillan.

Bortz, J. and Döring, N. (2006) *Forschungsmethoden und Evaluation für Human- und Sozialwissenschaftler*, 3rd edn. Berlin: Springer.

Bowen, G. A. (2009) 'Document Analysis as a Qualitative Research Method', *Qualitative Research Journal*, 9 (2): 27–40.

Braun, V. and Clarke, V. (2006) 'Using Thematic Analysis in Psychology', *Qualitative Research in Psychology*, 3(2): 77–101.

Brinkmann, S. and Kvale, S. (2018) *Doing Interviews*, 2nd edn. London: Sage.

British Educational Research Association [BERA] (2018) *Ethical Guidelines for Educational Research*, 4th edn. London. Available at: www.bera.ac.uk/researchers-resources/publications/ethical-guidelines-for-educational-research-2018 (accessed 4 November 2019).

Bryman, A. (1988) *Quantity and Quality in Social Research*. London: Unwin Hyman.

Bryman, A. (1992) 'Quantitative and Qualitative Research: Further Reflections on their Integration', in J. Brannen (ed.) *Mixing Methods: Quantitative and Qualitative Research*. Aldershot: Avebury. pp. 57–80.

Bryman, A. (2007) 'The Research Question in Social Research: What Is its Role?', *International Journal of Social Research Methodology*, 10(1): 5–20.

Bryman, A. (2016) *Social Research Methods*, 5th edn. Oxford: Oxford University Press.

Bude, H. (1989) 'Der Essay als Form der Darstellung sozialwissenschaftlicher Erkenntnisse', *Kölner Zeitschrift für Soziologie und Sozialpsychologie*, 41: 526–39.

Burra, T.A., Stergiopoulos, V. and Rourke, S.B. (2009) 'A Systematic Review of Cognitive Deficits in Homeless Adults: Implications for Service Delivery', *Canadian Journal of Psychiatry*, 54: 123–33.

Calvey, D. (2019) 'The Everyday World of Bouncers: A Rehabilitated Role for Covert Ethnography', *Qualitative Research*, 19(3): 247–62.

Campbell, D. (1986) 'Relabeling Internal and External Validity for Applied Social Sciences', in W.M.K. Trochim (ed.) *Advances in Quasi-Experimental Design Analysis*. San Francisco: Jossey-Bass. pp. 67–77.

Campbell, D.T. and Russo, M.J. (2001) *Social Measurement*. London: Sage.

Chandler, C. (2013) 'What is the Meaning of Impact in Relation to Research and Why Does it Matter? A View from Inside Academia', in P. Denicolo (ed.) *Achieving Impact in Research*. London: Sage. pp. 1–9.

Charmaz, K. (2014) *Constructing Grounded Theory: A Practical Guide through Qualitative Analysis*, 2nd edn. London: Sage.

Coffey, A. (2014) 'Analyzing Documents', in U. Flick (ed.) *The SAGE Handbook of Qualitative Data Analysis*. London: Sage. pp. 367–79.

Coldwell, C.M. and Bender, W.S. (2007) 'The Effectiveness of Assertive Community Treatment for Homeless Populations with Severe Mental Illness: A Meta-Analysis', *American Journal of Psychiatry*, 164: 393–9.

Corti, L. (2018) 'Data Collection in Secondary Analysis', in U. Flick (ed.) *The SAGE Handbook of Qualitative Data Collection*. London: Sage. pp. 164–82.

Corti, L. and Fielding, N. (2016) 'Opportunities from the Digital Revolution: Implications for Researching, Publishing, and Consuming Qualitative Research', *SAGE Open*, October–December, pp. 1–13.

Corti, L. and Wathan, J. (2017) 'Online Access to Quantitative Data Resources', in N. Fielding, R. Lee and G. Blank (eds) *The SAGE Handbook of Online Research Methods*. London: Sage. pp. 489–507.

Creswell, J.W. (1998) *Qualitative Inquiry and Research Design: Choosing Among Five Traditions*. Thousand Oaks, CA: Sage.

Creswell, J.W. (2003) *Research Design: Qualitative, Quantitative, and Mixed Methods Approaches*. Thousand Oaks, CA: Sage.

Creswell, J. W. (2015) *A Concise Introduction to Mixed Methods Research*. Thousand Oaks, CA: Sage.

Creswell, J.W. and Poth, C.N. (2017) *Research Design: Qualitative, Quantitative, and Mixed Methods Approaches*, 4th edn. Thousand Oaks, CA: Sage.

Creswell, J.W., Plano Clark, V.L., Gutman, M.L. and Hanson, W.E. (2003) 'Advanced Mixed Methods Research Design', in A. Tashakkori and C. Teddlie (eds) *Handbook of Mixed Methods in Social and Behavioral Research*. Thousand Oaks, CA: Sage. pp. 209–40.

Currie, C., Zanotti, C., Morgan, A., Currie, D., De Looze, M., Roberts, C. et al. (eds) (2012) *Social Determinants of Health and Well-being among Young People. Health Behaviour in School-aged Children (HBSC) Study: International Report from the 2009/2010 Survey*. Copenhagen: WHO Regional Office for Europe (Health Policy for Children and Adolescents, No. 6).

Dale, A., Arbor, S. and Procter, M. (1988) *Doing Secondary Analysis*. London: Unwin Hyman.

Denicolo, P. (ed.) (2013) *Achieving Impact in Research*. London: Sage.

Denscombe, M. (2007) *The Good Research Guide: For Small-Scale Social Research Projects*, 3rd edn. Maidenhead: McGraw-Hill.

Denzin, N.K. (1970/1989) *The Research Act* (1989, 3rd edn). Englewood Cliffs, NJ: Prentice Hall.

Denzin, N.K. (1988) *Interpretive Biography*. London: Sage.

Denzin, N.K. (2004) 'Reading Film: Using Photos and Video as Social Science Material', in U. Flick, E. v. Kardorff and I. Steinke (eds) *A Companion to Qualitative Research*. London: Sage. pp. 234–47.

Denzin, N. and Lincoln, Y.S. (2005) 'Introduction: The Discipline and Practice of Qualitative Research', in N. Denzin and Y.S. Lincoln (eds) *Handbook of Qualitative Research*, 3rd edn. London: Sage. pp. 1–32.

Department of Health (2001) *Research Governance Framework for Health and Social Care*. London: Department of Health.

Deutsche Forschungsgemeinschaft (DFG) (2013) *Vorschläge zur Sicherung guter wissenschaftlicher Praxis: Empfehlungen der Kommission 'Selbstkontrolle in der Wissenschaft'*, Denkschrift, 2nd edn. Weinheim: Wiley-VCH.

Diekmann, A. (2007) *Empirische Sozialforschung*. Reinbek: Rowohlt.

Drew, P. (1995) 'Conversation Analysis', in J.A. Smith, R. Harré and L. v. Langenhove (eds) *Rethinking Methods in Psychology*. London: Sage. pp. 64–79.

Edwards, A., Housley, W., Williams, M., Sloan, L. and Williams, M. (2013) 'Digital Social Research, Social Media and the Sociological Imagination: Surrogacy, Augmentation and Re-orientation', *International Journal of Social Research Methodology*, 16(3): 245–60.

Edwards, D. and Potter, J. (1992) *Discursive Psychology*. London: Sage.

Ellis, C., Adams, T.E. and Bochner, A.P. (2011) 'Autoethnography: An Overview', *Forum Qualitative Sozialforschung/Forum: Qualitative Social Research*, 12(1), Art. 10. Available in English at: https://www.dfg.de/download/pdf/dfg_im_profil/reden_stellungnahmen/download/empfehlung_wiss_praxis_1310.pdf (accessed 24 January 2020).

Ethik-Kodex (1993) 'Ethik-Kodex der Deutschen Gesellschaft für Soziologie und des Berufsverbandes Deutscher Soziologen', *DGS-Informationen*, 1/93: 13–19.

European Union (EU) (2018) *Ethics and data protection*. Available at: http://ec.europa.eu/research/participants/data/ref/h2020/grants_manual/hi/ethics/h2020_hi_ethics-data-protection_en.pdf (14 November) (accessed 21 April 2019).

Evans, J.R. and Mathur, A. (2005) 'The Value of Online Surveys', *Internet Research*, 15: 195–219.

Fielding, N.G. (2018) 'Combining Digital and Physical Data', in U. Flick (ed.) *The SAGE Handbook of Qualitative Data Collection*. London: Sage. pp. 584–98.

Fielding, N.G. and Fielding, J.L. (2000) 'Resistance and Adaptation to Criminal Identity: Using Secondary Analysis to Evaluate Classic Studies of Crime and Deviance', *Sociology*, 34(4): 671–89.

Fielding, N.G., Lee, R. and Blank, G. (eds) (2017) *The SAGE Handbook of Online Research Methods*, 2nd edn. London: Sage.

Fink, A. (2018) *Conducting Research Literature Reviews: From the Internet to Paper*, 5th edn. London: Sage.

Firsova, E. and Schmidt, M. (2017) *Chancen und Grenzen pädagogischer Interventionsarbeit zum Abbau antisemitischer Einstellungen bei Schülerinnen und Schülern*. Unpublished master's thesis, Freie Universität Berlin.

Fleck, C. (2004) 'Marie Jahoda', in U. Flick, E. v. Kardorff and I. Steinke (eds) *A Companion to Qualitative Research*. London: Sage. pp. 58–62.

Flick, U. (1992) 'Knowledge in the Definition of Social Situations: Actualization of Subjective Theories about Trust in Counseling', in M. v. Cranach, W. Doise and G. Mugny (eds) *Social Representations and the Social Bases of Knowledge*. Bern: Huber. pp. 64–8.

Flick, U. (1994) 'Social Representations and the Social Construction of Everyday Knowledge: Theoretical and Methodological Queries,' *Social Science Information*, 33(2): 179–97.

Flick, U. (ed.) (1998a) *Psychology of the Social: Representations in Knowledge and Language.* Cambridge: Cambridge University Press.

Flick, U. (1998b) 'The Social Construction of Individual and Public Health: Contributions of Social Representations Theory to a Social Science of Health', *Social Science Information*, 37: 639–62.

Flick, U. (2000a) 'Episodic Interviewing', in M. Bauer and G. Gaskell (eds) *Qualitative Researching with Text, Image and Sound: A Practical Handbook.* London: Sage. pp. 75–92.

Flick, U. (2000b) 'Qualitative Inquiries into Social Representations of Health', *Journal of Health Psychology*, 5: 309–18.

Flick, U. (2004a) 'Design and Process in Qualitative Research', in U. Flick, E. v. Kardorff and I. Steinke (eds) *A Companion to Qualitative Research.* London: Sage. pp. 146–52.

Flick, U. (2004b) 'Constructivism', in U. Flick, E. v. Kardorff and I. Steinke (eds) *A Companion to Qualitative Research.* London: Sage. pp. 88–94.

Flick, U. (ed.) (2014) *The SAGE Handbook of Qualitative Data Analysis.* London: Sage.

Flick, U. (2018a) *An Introduction to Qualitative Research*, 6th edn. London: Sage.

Flick, U. (2018b) *Designing Qualitative Research*, 2nd edn. London: Sage.

Flick, U. (ed.) (2018c) *The SAGE Qualitative Research Kit*, 2nd edn, 10 vols. London: Sage.

Flick, U. (ed.) (2018d) *The SAGE Handbook of Qualitative Data Collection.* London: Sage.

Flick, U. (2018e) *Doing Grounded Theory.* London: Sage.

Flick, U. (2018f) *Doing Triangulation and Mixed Methods.* London: Sage.

Flick, U. (2018g) 'Triangulation', in N.K. Denzin and Y.S. Lincoln (eds) *The SAGE Handbook of Qualitative Research*, 5th edn. London: Sage. pp. 444–61.

Flick, U. (2018h) *Managing Quality in Qualitative Research*, 2nd edn. London: Sage.

Flick, U. (forthcoming) *Doing Interview Research.* London: Sage.

Flick, U. and Röhnsch, G. (2007) 'Idealization and Neglect: Health Concepts of Homeless Adolescents', *Journal of Health Psychology*, 12: 737–50.

Flick, U. and Röhnsch, G. (2008) *Gesundheit und Krankheit auf der Straße: Vorstellungen und Erfahrungsweisen obdachloser Jugendlicher.* Weinheim: Juventa.

Flick, U., Fischer, C., Neuber, A., Walter, U. and Schwartz, F.W. (2003) 'Health in the Context of Being Old: Representations Held by Health Professionals', *Journal of Health Psychology*, 8(5): 539–56.

Flick, U., Garms-Homolová, V., Herrmann, W.J., Kuck, J. and Röhnsch, G. (2012) '"I Can't Prescribe Something Just Because Someone Asks for It ...": Using Mixed Methods in the Framework of Triangulation', *Journal of Mixed Methods Research*, 6(2): 97–110.

Flick, U., Garms-Homolová, V. and Röhnsch, G. (2010) '"When They Sleep, They Sleep": Daytime Activities and Sleep Disorders in Nursing Homes', *Journal of Health Psychology*, 15: 755–64.

Flick, U., Hoose, B. and Sitta, P. (1998) 'Gesundheit und Krankheit gleich Saúde & Doenca? Gesundheitsvorstellungen bei Frauen in Deutschland und Portugal', in U. Flick (ed.) *Wann fühlen wir uns gesund? Subjektive Vorstellungen von Gesundheit und Krankheit*. Weinheim: Juventa. pp. 141–59.

Flick, U., Kardorff, E. v. and Steinke, I. (eds) (2004) *A Companion to Qualitative Research*. London: Sage.

Frank, A. (1997) *The Wounded Storyteller: Body, Illness, and Ethics*. Chicago: University of Chicago Press.

Früh, W. (1991) *Inhaltsanalyse: Theorie und Praxis*, 3rd edn. München: Ölschläger.

Gaiser, T.J. and Schreiner, A.E. (2009) *A Guide to Conducting Online Research*. London: Sage.

Garms-Homolová, V., Flick, U. and Röhnsch, G. (2010) 'Sleep Disorders and Activities in Long Term Care Facilities: A Vicious Cycle?', *Journal of Health Psychology*, 15: 744–54.

Geertz, C. (1973) *The Interpretation of Cultures: Selected Essays*. New York: Basic.

Gergen, K.J. (1994) *Realities and Relationship: Soundings in Social Construction*. Cambridge, MA: Harvard University Press.

Gerhardt, U. (1988) 'Qualitative Sociology in the Federal Republic of Germany', *Qualitative Sociology*, 11: 29–43.

Gibbs, G. (2018) *Analyzing Qualitative Data*, 2nd edn. London: Sage.

Gill, F. and Elder, C. (2012) 'Data and Archives: The Internet as Site and Subject', *International Journal of Social Research Methodology*, 15(4): 271–9.

Gitelman, L. (ed.) (2013) *Raw Data is an Oxymoron*. Boston, MA: MIT Press.

Glaser, B.G. (1969) 'The Constant Comparative Method of Qualitative Analysis', in G.J. McCall and J.L. Simmons (eds) *Issues in Participant Observation*. Reading, MA: Addison-Wesley.

Glaser, B.G. (1978) *Theoretical Sensitivity*. Mill Valley, CA: University of California Press.

Glaser, B.G. and Strauss, A.L. (1965) *Awareness of Dying*. Chicago: Aldine.

Glaser, B.G. and Strauss, A.L. (1967) *The Discovery of Grounded Theory: Strategies for Qualitative Research*. New York: Aldine.

Greene, J.C. (2007) *Mixed Methods in Social Inquiry*. San Francisco: Jossey-Bass.

Greenhalgh, T. (2006) *How to Read a Paper: The Basics of Evidence-Based Medicine*. Oxford: Blackwell-Wiley.

Guba, E.G. and Lincoln, Y.S. (1989) *Fourth Generation Evaluation*. Newbury Park, CA: Sage.

Guggenmoos-Holzmann, I., Bloomfield, K., Brenner, H. and Flick, U. (eds) (1995) *Quality of Life and Health: Concepts, Methods and Applications*. Berlin: Blackwell Science.

Guiney Yallop, J.J., Lopez de Vallejo, I. and Wright, P.R. (2009) 'Editorial: Overview of the Performative Social Science Special Issue', *Forum Qualitative Sozialforschung/ Forum: Qualitative Social Research*, 9(2), Art. 64. Available at: http://nbn-resolving.de/urn:nbn:de:0114-fqs0802649 (accessed 28 November 2014).

Hakim, C. (1982) *Secondary Analysis in Social Research: A Guide to Data Sources and Method Examples*. London: George Allen & Uwin.

Halkier, B. (2010) 'Focus Groups as Social Enactments: Integrating Interaction and Content in the Analysis of Focus Group Data', *Qualitative Research*, 10(1): 71–89.

Hammersley, M. and Atkinson, P. (1995) *Ethnography: Principles in Practice*, 2nd edn. London: Routledge.

Hannes, K. (2011) 'Critical Appraisal of Qualitative Research', in J. Noyes, A. Booth, K. Hannes, A. Harden, J. Harris, S. Lewin and C. Lockwood (eds) *Supplementary Guidance for Inclusion of Qualitative Research in Cochrane Systematic Reviews of Interventions*. Version 1 (updated August 2011). Cochrane Collaboration Qualitative Methods Group, Chapter 4. Available at: http://cqrmg.cochrane.org/supplemental-handbook-guidance (accessed 1 September 2014).

Harré, R. (1998) 'The Epistemology of Social Representations', in U. Flick (ed.) *Psychology of the Social: Representations in Knowledge and Language*. Cambridge: Cambridge University Press. pp. 129–37.

Hart, C. (1998) *Doing a Literature Review*. London: Sage.

Hart, C. (2001) *Doing a Literature Search*. London: Sage.

Hart, T. (2014) 'Technologies for Conducting an Online Ethnography of Communication: The Case of Eloqi', in S. Hai-Jew (ed.) *Enhancing Qualitative and Mixed Methods Research with Technology*. Hershey, PA: IGI Global. pp. 105–24.

Hart, T. (2017) 'Online Ethnography', in J. Matthes, C. S. Davis, and R. F. Potter (eds) *The International Encyclopedia of Communication Research Methods*. doi: 10.1002/9781118901731.iecrm0172

Hermanns, H. (1995) 'Narrative Interviews', in U. Flick, E. v. Kardorff, H. Keupp, L. v. Rosenstiel and S. Wolff (eds) *Handbuch Qualitative Sozialforschung*, 2nd edn. Munich: Psychologie Verlags Union. pp. 182–5.

Herrmann, W.J. and Flick, U. (2012) 'Nursing Home Residents' Psychological Barriers to Sleeping Well: A Qualitative Study', *Family Practice*, 29(1): 482–7.

Hewson, C., Yule, P., Laurent, D. and Vogel, C. (2003) *Internet Research Methods: A Practical Guide for the Social and Behavioural Sciences*. London: Sage.

Higgins, J.P.T. and Green, S. (eds) (2011) *Cochrane Handbook for Systematic Reviews of Interventions*, Version 5.1.0 (updated March 2011), the Cochrane Collaboration. Available at: www.cochrane-handbook.org (accessed 1 September 2014).

Hine, C. (2000) *Virtual Ethnography*. London: Sage.

Hochschild, A.R. (1983) *The Managed Heart*. Berkeley, CA: University of California Press.

Hoinville, G., Jowell, R. and associates (1985) *Survey Research Practice*. Aldershot: Gower.

Hollingshead, A.B. and Redlich, F.C. (1958) *Social Class and Mental Illness: A Community Sample*. New York: Wiley.

Hopf, C. (2004) 'Research Ethics and Qualitative Research: An Overview', in U. Flick, E. v. Kardorff and I. Steinke (eds) *A Companion to Qualitative Research*. London: Sage. pp. 334–9.

Humphreys, L. (1975) *Tearoom Trade: Impersonal Sex in Public Places* (enlarged edn). New York: Aldine Transaction.

Hurrelmann, K., Klocke, A., Melzer, W. and Ravens-Sieberer, U. (2003) *Konzept und ausgewählte Ergebnisse der Studie*. Available at: www.hbsc-germany.de/pdf/artikel_ hurrelmann_klocke_melzer_urs.pdf (accessed 30 June 2008).

Irwin, S., Bornat, J. and Winterton, M. (2012) 'Timescapes Secondary Analysis: Comparison, Context and Working across Data Sets', *Qualitative Research*, 12(1): 66–80.

Jahoda, M. (1995) 'Jahoda, M., Lazarsfeld, P. and Zeisel, H. (1933) Die Arbeitslosen von Marienthal', in U. Flick, E. v. Kardorff, H. Keupp, L. v. Rosenstiel and S. Wolff (eds) *Handbuch Qualitative Sozialforschung*, 2nd edn. München: Psychologie Verlags Union. pp. 119–22.

Jahoda, M., Lazarsfeld, P.F. and Zeisel, H. (1933/1971) *Marienthal: The Sociology of an Unemployed Community*. Chicago: Aldine-Atherton.

Jörgensen, D.L. (1989) *Participant Observation: A Methodology for Human Studies*. London: Sage.

Kasperiuniene, J. and Zydziunaite, V. (2019) 'A Systematic Literature Review on Professional Identity Construction in Social Media', *SAGE Open*. https://doi. org/10.1177/2158244019828847

Kaulmann, A. (2017) *Konsequenzen einer Mitgliedschaft in einem One-Percenter-Motorcycle-Club for die Eltern-Kind-Beziehung und die Sozialisation des Kindes*. Unpublished master's thesis, Freie Universität Berlin.

Kelle, U. (1994) *Empirisch begründete Theoriebildung: Zur Logik und Methodologie interpretativer Sozialforschung*. Weinheim: Deutscher Studienverlag.

Kelle, U. and Erzberger, C. (2004) 'Quantitative and Qualitative Methods: No Confrontation', in U. Flick, E. v. Kardorff and I. Steinke (eds) *A Companion to Qualitative Research*. London: Sage. pp. 172–7.

Kelle, U. and Kluge, S. (2010) *Vom Einzelfall zum Typus: Fallvergleich und Fallkontrastierung in der qualitativen Sozialforschung*. Wiesbaden: VS-Verlag.

Kelly, K. and Caputo, T. (2007) 'Health and Street/Homeless Youth', *Journal of Health Psychology*, 12: 726–36.

Kendall, L. (1999) 'Recontextualising Cyberspace: Methodological Considerations for On-Line Research', in S. Jones (ed.) *Doing Internet Research: Critical Issues and Methods for Examining the Net*. London: Sage. pp. 57–74.

Khazal, J. (2019) *Welchen Einfluss haben Fluchterfahrungen auf die Emanzipation der syrischen Frau in Deutschland?* Unpublished thesis, Freie Universität Berlin.

Kirk, J.L. and Miller, M. (1986) *Reliability and Validity in Qualitative Research*. Beverley Hills, CA: Sage.

Kitchin, R. (2014) 'Big Data, New Epistemologies and Paradigm Shifts', *Big Data & Society*. https://doi.org/10.1177/2053951714528481

Kitchin, R. (2017) 'Big Data – Hype or Revolution?', in L. Sloan and A. Quan-Haase (eds) *The SAGE Handbook of Social Media Research Methods*. London: Sage. pp. 27–39.

Knoblauch, H., Schnettler, B., Raab, J. and Soeffner, H.-G. (eds) (2006) *Video Analysis: Methodology and Methods*. Frankfurt: Lang.

Knoblauch, H., Schnettler, B. and Tuma, R. (2018) 'Videography', in U. Flick (ed.) *The SAGE Handbook of Qualitative Data Collection*. London: Sage. pp. 362–77.

Knoblauch, H., Tuma, R. and Schnettler, B. (2014) 'Video Analysis and Videography', in U. Flick (ed.) *The SAGE Handbook of Qualitative Data Analysis*. London: Sage. pp. 435–49.

Kozinets, R.V. (2010) *Netnography: Doing Ethnographic Research Online*. London: Sage.

Kozinets, R.V., Dalbec, P.-Y. and Earley, A. (2014) 'Netnographic Analysis: Understanding Culture through Social Media Data', in U. Flick (ed.) *The SAGE Handbook of Qualitative Data Analysis*. London: Sage. pp. 262–74.

Krippendorf, K. (2012) *Content Analysis: An Introduction to its Methodology*, 3rd edn. London: Sage.

Kromrey, H. (2006) *Empirische Sozialforschung: Modelle und Methoden der standardisierten Datenerhebung und Datenauswertung*. Opladen: Leske and Budrich/UTB.

Kuhn, T.S. (1970) *The Structure of Scientific Revolutions*, 2nd edn. Chicago: University of Chicago Press (first published 1965).

Kvale, S. (2007) *Doing Interviews*. London: Sage.

Lakatos, I. (1980) *The Methodology of Scientific Research Programmes: Philosophical Papers Volume 1*. Cambridge: Cambridge University Press.

Largan, C. and Morris, T. (2019) *Qualitative Secondary Research: A Step-by-Step Guide*. London: Sage.

Laswell, H.D. (1938) 'A Provisional Classification of Symbol Data', *Psychiatry*, 1: 197–204.

Lechner, B. (2019) *Stigmatisierung und psychische Erkrankung in der landesparlamentarischen Debatte zum bayerischen Psychisch-Kranken-Hilfe-Gesetz*. Unpublished master's thesis, Freie Universität Berlin.

Legewie, H. (1987) 'Interpretation und Validierung biographischer Interviews', in G. Jüttemann and H. Thomae (eds) *Biographie und Psychologie*. Berlin: Springer. pp. 138–50.

Liamputtong, P. (2007) *Researching the Vulnerable: A Guide to Sensitive Research Methods.* Thousand Oaks, CA: Sage.

Lillis, G. (2002) *Delivering Results in Qualitative Market Research,* vol. 7 of the *Qualitative Market Research Kit.* London: Sage.

Lincoln, Y.S. and Guba, E.G. (1985) *Naturalistic Inquiry.* London: Sage.

Lofland, J.H. (1974) 'Styles of Reporting Qualitative Field Research', *American Sociologist,* 9: 101–11.

Lofland, J. (1976) *Doing Social Life: The Qualitative Study of Human Interaction in Natural Settings.* New York: Wiley.

Lofland, J. and Lofland, L.H. (1984) *Analyzing Social Settings,* 2nd edn. Belmont, CA: Wadsworth.

Lucius-Hoene, G. and Deppermann, A. (2002) *Rekonstruktion narrativer Identität: Ein Arbeitsbuch zur Analyse narrativer Interviews.* Opladen: Leske and Budrich.

Lüders, C. (1995) 'Von der Teilnehmenden Beobachtung zur ethnographischen Beschreibung: Ein Literaturbericht', in E. König and P. Zedler (eds) *Bilanz qualitativer Forschung,* vol. 1. Weinheim: Deutscher Studienverlag. pp. 311–42.

Lüders, C. (2004a) 'Field Observation and Ethnography', in U. Flick, E. v. Kardorff and I. Steinke (eds) *A Companion to Qualitative Research.* London: Sage. pp. 222–30.

Lüders, C. (2004b) 'The Challenges of Qualitative Research', in U. Flick, E. v. Kardorff and I. Steinke (eds) *A Companion to Qualitative Research.* London: Sage. pp. 359–64.

Lüders, C. (2006) 'Qualitative Daten als Grundlage der Politikberatung', in U. Flick (ed.) *Qualitative Evaluationsforschung: Konzepte, Methoden, Umsetzungen.* Reinbek: Rowohlt. pp. 444–62.

Mallinson, S. (2002) 'Listening to Respondents: A Qualitative Assessment of the Short-Form 36 Health Status Questionnaire', *Social Science and Medicine,* 54: 11–21.

Mann, C. and Stewart, F. (2000) *Internet Communication and Qualitative Research: A Handbook for Researching Online.* London: Sage.

Markham, A.M. (2004) 'The Internet as Research Context Research', in C. Seale, G. Gobo, J. Gubrium and D. Silverman (eds) *Qualitative Research Practice.* London: Sage. pp. 358–74.

Marotzki, W., Holze, J. and Verständig, D. (2014) 'Analyzing Virtual Data', in U. Flick (ed.) *The SAGE Handbook of Qualitative Data Analysis.* London: Sage. pp. 450–64.

Marshall, C. and Rossman, G.B. (2006) *Designing Qualitative Research,* 4th edn. Thousand Oaks, CA: Sage.

Matt, E. (2004) 'The Presentation of Qualitative Research', in U. Flick, E. v. Kardorff and I. Steinke (eds) *A Companion to Qualitative Research.* London: Sage. pp. 326–30.

Maxwell, J.A. (2005) *Qualitative Research Design: An Interactive Approach,* 2nd edn. Thousand Oaks, CA: Sage.

Maxwell, J. and Chmiel, M. (2014) 'Notes Toward a Theory of Qualitative Data Analysis', in U. Flick (ed.) *The SAGE Handbook of Qualitative Data Analysis*. London: Sage. pp. 21–34.

May, T. (2001) *Social Research: Issues, Methods and Process*. Maidenhead: Open University Press.

Mayr, P. and Weller, K. (2017) 'Think Before You Collect: Setting up a Data Collection Approach for Social Media Studies', in L. Sloan and A. Quan-Haase (eds) *The SAGE Handbook of Social Media Research Methods*. London: Sage. pp. 107–24.

Mayring, P. (1983) *Qualitative Inhaltsanalyse: Grundlagen und Techniken*. Weinheim: Deutscher Studien Verlag.

Mayring, P. (2000) 'Qualitative Content Analysis' [28 paragraphs], *Forum Qualitative Sozialforschung / Forum: Qualitative Social Research*, 1(2): Art. 20, http://nbn-resolving. de/urn:nbn:de:0114-fqs0002204 (last accessed 13 August 2013).

McCay-Peet, L. and Quan-Haase, A. (2017) 'What is Social Media and What Questions Can Social Media Research Help Us Answer?', in L. Sloan and A. Quan-Haase (eds) *The SAGE Handbook of Social Media Research Methods*. London: Sage. pp. 13–26.

Mensching, A. (2006) 'Zwischen Überforderung und Banalisierung: zu den Schwierigkeiten der Vermittlungsarbeit im Rahmen qualitativer Evaluationsforschung', in U. Flick (ed.) *Qualitative Evaluationsforschung: Konzepte, Methoden, Umsetzungen*. Reinbek: Rowohlt. pp. 339–60.

Mertens, D.M. (2014) 'Ethical Use of Qualitative Data and Findings', in U. Flick (ed.) *The SAGE Handbook of Qualitative Data Analysis*. London: Sage. pp. 510–23.

Mertens, D. and Ginsberg, P.E. (eds) (2009) *Handbook of Research Ethics*. London: Sage.

Merton, R.K. and Kendall, P.L. (1946) 'The Focused Interview', *American Journal of Sociology*, 51: 541–57.

Meuser, M. (2003) 'Interpretatives Paradigma', in R. Bohnsack, W. Marotzki and M. Meuser (eds) *Hauptbegriffe Qualitativer Sozialforschung: Ein Wörterbuch*. Opladen: Leske + Budrich. pp. 92–4.

Meuser, M. and Nagel, U. (2002) 'Experteninterviews – vielfach erprobt, wenig bedacht: Ein Beitrag zur qualitativen Methodendiskussion', in A. Bogner, B. Littig and W. Menz (eds) *Das Experteninterview*. Opladen: Leske & Budrich. pp. 71–95.

Meuser, M. and Nagel, U. (2009) 'The Expert Interview and Changes in Knowledge Production', in A. Bogner, B. Littig and W. Menz (eds) *Interviewing Experts*. Basingstoke: Palgrave Macmillan. pp. 17–42.

Mikos, L. (2014) 'Analysis of Film', in U. Flick (ed.) *The SAGE Handbook of Qualitative Data Analysis*. London: Sage. pp. 409–23.

Mikos, L. (2018) 'Collecting Media Data: TV and Film Studies', in U. Flick (ed.) *The SAGE Handbook of Qualitative Data Collection*. London: Sage. pp. 412–25.

Miles, M.B. and Huberman, A.M. (1994) *Qualitative Data Analysis: A Sourcebook of New Methods*, 2nd edn. Newbury Park, CA: Sage.

Miles, M.B., Huberman, A.M. and Saldana, J. (2018) *Qualitative Data Analysis: An Expanded Sourcebook*, 4th edn. Thousand Oaks, CA: Sage.

Mishler, E.G. (1990) 'Validation in Inquiry-Guided Research: The Role of Exemplars in Narrative Studies', *Harvard Educational Review*, 60: 415–42.

Morgan, D.L. and Hoffman, K. (2018) 'Focus Groups', in U. Flick (ed.), *The SAGE Handbook of Qualitative Data Collection*. London: Sage, pp. 250–63.

Morse, J.M. (1998) 'Designing Funded Qualitative Research', in N. Denzin and Y.S. Lincoln (eds) *Strategies of Qualitative Research*. London: Sage. pp. 56–85.

Morse, J.M. (2001) 'Qualitative Verification: Strategies for Extending the Findings of a Research Project', in J.M. Morse, J. Swanson and A. Kuzel (eds) *The Nature of Evidence in Qualitative Inquiry*. Newbury Park, CA: Sage. pp. 203–21.

Morse, J.M., Swanson, J. and Kuzel, A.J. (eds) (2001) *The Nature of Qualitative Evidence in Qualitative Inquiry*. Thousand Oaks, CA: Sage.

Murphy, E. and Dingwall, R. (2001) 'The Ethics of Ethnography', in P. Atkinson, A. Coffey, S. Delamont, J. Lofland and L. Lofland (eds) *Handbook of Ethnography*. London: Sage. pp. 339–51.

Murray, M. (2014) 'Implementation: Putting Analyses into Practice', in U. Flick (ed.) *The SAGE Handbook of Qualitative Data Analysis*. London: Sage. pp. 585–99.

Neuman, W.L. (2014) *Social Research Methods: Qualitative and Quantitative Approaches*, 7th edn. Harlow: Pearson.

Neville, C. (2010) *Complete Guide to Referencing and Avoiding Plagiarism*. Maidenhead: Open University Press.

Panke, D. (2018) *Research Design and Method Selection: Making Good Choices in the Social Sciences*. London: Sage.

Papacharissi, Z. (2015) 'We Have Always Been Social', *Social Media + Society*, April–June: 1–2.

Patton, M.Q. (2002) *Qualitative Evaluation and Research Methods*, 3rd edn. London: Sage.

Pohlenz, P., Hagenmüller, J.-P. and Niedermeier, F. (2009) 'Ein Online-Panel zur Analyse von Studienbiographien: Qualitätssicherung von Lehre und Studium durch webbasierte Sozialforschung', in N. Jackob, H. Schoen and T. Zerback (eds) *Sozialforschung im Internet: Methodologie und Praxis der Online-Befragung*. Wiesbaden: VS-Verlag. pp. 233–44.

Polk, D. (2017) 'Symbolic Interactionism', in M. Allan (ed.) *The SAGE Encyclopedia of Communication Research Methods*. Thousand Oaks, CA: Sage. pp. 1739–43.

Popper, K. (1971) *Logik der Forschung*. Tübingen: Mohr (first published 1934).

Porst, R. (2014) *Fragebogen*. Wiesbaden: Springer Fachmedien.

Potter, J. and Shaw, C. (2018) 'The Virtues of Naturalistic Data', in U. Flick (ed.) *The SAGE Handbook of Qualitative Data Collection*. London: Sage. pp. 182–99.

Potter, J. and Wetherell, M. (1998) 'Social Representations, Discourse Analysis, and Racism', in U. Flick (ed.) *Psychology of the Social: Representations in Knowledge and Language*. Cambridge: Cambridge University Press. pp. 177–200.

Procter, R., Vis, F. and Voss, A. (2013) 'Reading the Riots on Twitter: Methodological Innovation for the Analysis of Big Data', *International Journal of Social Research Methodology*, 16(3): 197–214.

Punch, K. (1998) *Introduction to Social Research*. London: Sage.

Radey, M. (2010) 'Secondary Data Analysis Studies', in B. Thyer (ed.) *The Handbook of Social Work Research Methods*. London: Sage. pp. 163–82.

Ragin, C.C. (1994) *Constructing Social Research*. Thousand Oaks, CA: Pine Forge.

Ragin, C.C. and Becker, H.S. (eds) (1992) *What is a Case? Exploring the Foundations of Social Inquiry*. Cambridge: Cambridge University Press.

Rapley, T. (2018) *Doing Conversation, Discourse and Document Analysis*, 2nd edn. London: Sage.

Rew, L., Fouladi, R. T., Land, L. and Wong, Y. J. (2007) 'Outcomes of a Brief Sexual Health Intervention for Homeless Youth', *Journal of Health Psychology*, 12(5): 818–32.

Richter, M. (2003) 'Anlage und Methode des Jugendgesundheitssurveys', in K. Hurrelmann, A. Klocke, W. Melzer and U. Ravens-Sieberer (eds) *Jugendgesundheits- survey: Internationale Vergleichsstudie im Auftrag der Weltgesundheitsorganisation WHO*. Weinheim: Juventa. pp. 9–18.

Richter, M., Hurrelmann, K., Klocke, A., Melzer, W. and Ravens-Sieberer, U. (eds) (2008) *Gesundheit, Ungleichheit und jugendliche Lebenswelten: Ergebnisse der zweiten internationalen Vergleichsstudie im Auftrag der Weltgesundheitsorganisation WHO*. Weinheim: Juventa.

Riemann, G. and Schütze, F. (1987) 'Trajectory as a Basic Theoretical Concept for Analyzing Suffering and Disorderly Social Processes', in D. Maines (ed.) *Social Organization and Social Process: Essays in Honor of Anselm Strauss*. New York: Aldine de Gruyter. pp. 333–57.

RIN (2010) *If You Build It, Will They Come? How Researchers Perceive and Use Web 2.0*. Available at: www.rin.ac.uk/our-work/communicating-and-disseminating-research/ use-and-relevance-web-20-researchers (accessed 21 August 2010).

Rosenberg, M. (1968) *The Logic of Survey Analysis*. New York: Basic.

Rosenthal, G. and Fischer-Rosenthal, W. (2004) 'The Analysis of Biographical-Narrative Interviews', in U. Flick, E. v. Kardorff and I. Steinke (eds) *A Companion to Qualitative Research*. London: Sage. pp. 259–65.

Sahle, R. (1987) *Gabe, Almosen, Hilfe*. Opladen: Westdeutscher Verlag.

Saldana, J. (2013) *The Coding Manual for Qualitative Researchers*, 2nd edn. London: Sage.

Salmons, J. (2010) *Online Interviews in Real Time*. London: Sage.

Sandberg, J. and Alvesson, M. (2010) 'Ways of Constructing Research Questions: Gap-spotting or Problematization?', *Organization*, 18(1): 23–44.

Sandelowski, M. and Leeman, J. (2012) 'Writing Usable Qualitative Health Research Findings', *Qualitative Health Research*, 10: 1404–13.

Savage, M. (2005) 'Revisiting Classic Qualitative Studies' [43 paragraphs]. *Forum Qualitative Sozialforschung / Forum: Qualitative Social Research*, 6(1), Art. 31, http://nbn-resolving.de/:nbn:de:0114-fqs0501312.

Schnell, M.W. and Heinritz, C. (2006) *Forschungsethik: Ein Grundlagen- und Arbeitsbuch mit Beispielen aus der Gesundheits- und Pflegewissenschaft*. Bern: Huber.

Schnell, R., Hill, P.B. and Esser, E. (2008) *Methoden der empirischen Sozialforschung*. München: Oldenbourg.

Schreier, M. (2012) *Qualitative Content Analysis in Practice*. London: Sage.

Schreier, M. (2014) 'Qualitative Content Analysis', in U. Flick (ed.) *The SAGE Handbook of Qualitative Data Analysis*. London: Sage. pp. 170–83.

Schutt, R. (2014) *Investigating the Social World: The Process and Practice of Research*, 8th edn. London: Sage.

Schütz, A. (1962) *Collected Papers, Vols I and II*. The Hague: Nijhoff.

Schütze, F. (1983) 'Biographieforschung und Narratives Interview', *Neue Praxis*, 3: 283–93.

Scott, J. (1990) *A Matter of Record: Documentary Sources in Social Research*. Cambridge: Polity Press.

Seale, C. (1999) *The Quality of Qualitative Research*. London: Sage.

Shaw, I., Ramatowski, A. and Ruckdeschel, R. (2013) 'Patterns, Designs and Developments in Qualitative Research in Social Work: A Research Note', *Qualitative Social Work*, 12(6): 732–49.

Silverman, D. (2001) *Interpreting Qualitative Data: Methods for Analysing Talk, Text and Interaction*, 2nd edn. London: Sage.

Simons, H. (2009) *Case Study Research in Practice*. London: Sage.

Slater, J. R. (2017) 'Social Constructionism', in M. Allan (ed.) *The SAGE Encyclopedia of Communication Research Methods*. Thousand Oaks, CA: Sage. pp. 1624–8.

Sörensson, E. and Kalman, H. (2018) 'Care and Concern in the Research Process: Meeting Ethical and Epistemological Challenges through Multiple Engagements and Dialogue with Research Subjects', *Qualitative Research*, 18(6): 706-21.

Spradley, J.P. (1980) *Participant Observation*. New York: Rinehart & Winston.

Sprenger, A. (1989) 'Teilnehmende Beobachtung in prekären Handlungssituationen: Das Beispiel Intensivstation', in R. Aster, H. Merkens and M. Repp (eds)

Teilnehmende Beobachtung: Werkstattberichte und methodologische Reflexionen. Frankfurt: Campus. pp. 35–56.

Stake, R.E. (1995) *The Art of Case Study Research.* London: Sage.

Stark, K. and Guggenmoos-Holzmann, I. (2003) 'Wissenschaftliche Ergebnisse deuten und nutzen', in F.W. Schwartz (ed.) *Das Public Health Buch.* Heidelberg: Elsevier, Urban and Fischer. pp. 393–418.

Stewart, B. (2017) 'Twitter as Method: Using Twitter as a Tool to Conduct Research', in L. Sloan and A. Quan-Haase (eds), *The SAGE Handbook of Social Media Research Methods.* London: Sage, pp. 251–65.

Strauss, A.L. (1987) *Qualitative Analysis for Social Scientists.* Cambridge: Cambridge University Press.

Strauss, A.L. and Corbin, J. (1990/1998/2008) *Basics of Qualitative Research* (2nd edn 1998, 3rd edn 2008). London: Sage.

Stryker, S. (1976) 'Die Theorie des Symbolischen Interaktionismus', in M. Auwärter, E. Kirsch and K. Schröter (eds) *Seminar: Kommunikation, Interaktion, Identität.* Frankfurt: Suhrkamp. pp. 257–74.

Sue, V.M. (2007) *Conducting Online Surveys.* London: Sage.

Tashakkori, A. and Teddlie, C. (eds) (2003a) *Handbook of Mixed Methods in Social and Behavioral Research.* Thousand Oaks, CA: Sage.

Tashakkori, A. and Teddlie, C. (2003b) 'Major Issues and Controversies in the Use of Mixed Methods in Social and Behavioral Research', in A. Tashakkori and C. Teddlie (eds) *Handbook of Mixed Methods in Social and Behavioral Research.* Thousand Oaks, CA: Sage. pp. 3–50.

Tashakkori, A. and Teddlie, C. (eds) (2010) *Handbook of Mixed Methods in Social and Behavioral Research,* 2nd edn. Thousand Oaks, CA: Sage.

Taylor, S. (2012) '"One Participant Said …": The Implications of Quotations from Biographical Talk', *Qualitative Research,* 12(4): 388–401.

Thomson, H., Hoskins, R., Petticrew, M., Ogilvie, D., Craig, N., Quinn, T. and Lindsay, G. (2004) 'Evaluating the Health Effects of Social Interventions', *BMJ,* 328: 282–5.

Thornberg, R. and Charmaz, K. (2014) 'Grounded Theory and Theoretical Coding', in U. Flick (ed.) *The SAGE Handbook of Qualitative Data Analysis.* London: Sage. pp. 153–69.

Tiidenberg, K. (2018) 'Ethics in Digital Research', in U. Flick (ed.) *The SAGE Handbook of Qualitative Data Collection.* London: Sage. pp. 466–81.

Tight, M. (2019) *Documentary Research in the Social Sciences.* London: Sage.

Timulak, L. (2014) 'Qualitative Meta-Analysis', in U. Flick (ed.) *The SAGE Handbook of Qualitative Data Analysis.* London: Sage. pp. 481–95.

Tinati, R., Halford, S., Carr, L. and Pope, C. (2014) 'Big Data: Methodological Challenges and Approaches for Sociological Analysis', *Sociology,* 48(4): 663–81.

Toerien, M. (2014) 'Conversations and Conversation Analysis', in U. Flick (ed.) *The SAGE Handbook of Qualitative Data Analysis*. London: Sage. pp. 327–40.

Tracy, S.J. (2010) 'Qualitative Quality: Eight "Big-Tent" Criteria for Excellent Qualitative Research', *Qualitative Inquiry*, 16: 837–51.

Tufte, E.R. (1990) *Envisioning Information*. Cheshire, CT: Graphics.

Tufte, E.R. (2001) *Visual Display of Quantitative Information*. Cheshire, CT: Graphics.

UCL (2018) *Guidance Paper for Researchers*. Available at: www.ucl.ac.uk/legal-services/ucl-general-data-protection-regulation-gdpr/guidance-notices-ucl-staff/guidance-researchers (accessed 22 April 2019).

UKRI (2018) *GDPR and Research: An Overview for Researchers – What is GDPR?* Available at: www.ukri.org/files/about/policy/ukri-gdpr-faqs-pdf (accessed 23 April 2019).

University Library (2009) 'Gray Literature', California State University, Long Beach. Available at: www.csulb.edu/library/subj/gray_literature (accessed 17 August 2010).

Van Maanen, J. (1988) *Tales of the Field: On Writing Ethnography*. Chicago: University of Chicago Press.

Wächter, F. (2019) *Bedeutungen von Spiritualität im Hospiz – Eine Rekonstruktion subjektiver Sichtweisen auf Spiritualität unter Pflegekräften und Sozialarbeiterinnen*. Unpublished master's thesis, Freie Universität Berlin.

Walter, U., Flick, U., Fischer, C., Neuber, A. and Schwartz, F.W. (2006) *Alt und gesund? Altersbilder und Präventionskonzepte in der ärztlichen und pflegerischen Praxis*. Wiesbaden: VS-Verlag.

Wästerfors, D., Åkerström, M. and Jacobsson, K. (2014) 'Re-analysis of Qualitative Data', in U. Flick (ed.) *The SAGE Handbook of Qualitative Data Analysis*. London: Sage. pp. 467–80.

Weingart, P. (2001) *Die Stunde der Wahrheit? Zum Verhältnis der Wissenschaft zu Politik, Wirtschaft und Medien in der Wissensgesellschaft*. Weilerswist: Velbrück.

Weischer, C. (2007) *Sozialforschung: Theorie und Praxis*. Konstanz: UVK-UTB.

Wiedemann, P.M. (1995) 'Gegenstandsnahe Theoriebildung', in U. Flick, E. v. Kardorff, H. Keupp, L. v. Rosenstiel and S. Wolff (eds) *Handbuch Qualitative Sozialforschung*, 2nd edn. Munich: Psychologie Verlags Union. pp. 440–5.

Willig, C. (2003) *Introducing Qualitative Research in Psychology: Adventures in Theory and Method*. Milton Keynes: Open University Press.

Willig, C. (2014a) 'Discourses and Discourse Analysis', in U. Flick (ed.) *The SAGE Handbook of Qualitative Data Analysis*. London: Sage. pp. 341–53.

Willig, C. (2014b) 'Interpretation and Analysis', in U. Flick (ed.) *The SAGE Handbook of Qualitative Data Analysis*. London: Sage. pp. 136–50.

Wilson, T.P. (1970) 'Normative and Interpretive Paradigms in Sociology', in J.D. Douglas (ed.) *Understanding Everyday Life: Toward the Reconstruction of Sociological Knowledge*. London: Routledge & Paul. pp. 52–79.

Wilson, T.P. (1982) 'Quantitative "oder" qualitative Methoden in der Sozialforschung', *Kölner Zeitschrift für Soziologie und Sozialpsychologie*, 34: 487–508.

Wolff, S. (1987) 'Rapport und Report: Über einige Probleme bei der Erstellung plausibler ethnographischer Texte', in W. von der Ohe (ed.) *Kulturanthropologie: Beiträge zum Neubeginn einer Disziplin*. Berlin: Reimer. pp. 333–64.

Wolff, S. (2004) 'Analysis of Documents and Records', in U. Flick, E. v. Kardorff and I. Steinke (eds) *A Companion to Qualitative Research*. London: Sage. pp. 284–90.

World Health Organization (WHO) (1986) *Ottawa Charter for Health Promotion*. First International Conference on Health Promotion, Ottawa. Available at: www.who.int/healthpromotion/conferences/previous/ottawa/en/ (accessed 28 November 2014).

Worthington, H. and Holloway, P. (1997) 'Making Sense of Statistics', *International Journal of Health Education*, 35 (3): 76–80.

NAME INDEX

European Union (EU), 53–55
Evans, J.R., 241

Fielding, J., 202
Fielding, N.G., 16, 191, 201, 202–203,
  214–215
Firsova, E., 113
Fischer-Rosenthal, W., 287
Fleck, C., 62
Flick, U., 10, 29, 46, 63–64, 67, 68, 69, 70,
  83, 116, 118, 127, 135, 136, 168, 186,
  188, 190, 191, 192–193, 213, 236, 237,
  238, 256–257, 280, 284, 291, 308, 309,
  310, 333–335, 336–337, 340
Florian, M., 146
Frank, A., 333
Früh, W., 262

Gaiser, T.J., 51–52, 148
Garms-Homolová, V., 191
Geertz, C., 333
Gergen, K.J., 30–31
Gerhardt, U., 292, 314
Gill, F., 205–206
Gitelman, L., 215
Glaser, B.G., 63, 64, 116, 118,
  144–145, 166, 171, 280, 282–283,
  287, 291, 308, 313–314, 338–339
Green, S., 337
Greene, J.C., 293
Greenhalgh, T., 331–332
Guba, E.G, 310–312, 313
Guggenmoos-Holzmann, I.,
  256, 331
Guiney Yallop, J.J., 338

Hakim, C., 202
Halkier, B., 289
Hammersley, M., 25, 250
Hannes, K., 337
Harré, R., 289–290
Hart, C., 94
Hart, T., 211, 252
Heinritz, C., 40–41, 42, 49, 56
Hermanns, H., 234–235
Herrmann, W.J., 46
Hewson, C., 148, 241, 242

Higgins, J.P.T., 337
Hine, C., 251
Hochschild, A.R., 63, 64
Hoffman, K., 239
Hoinville, G., 160–161
Hollingshead, A.B., 62, 64, 73
Holloway, P., 272
Hoose, B., 63–64, 136
Hopf, C., 50
Huberman, A.M., 170, 189, 191
Humphreys, L., 42–43
Hurrelmann, K., 8, 65–66

Irwin, S., 202

Jahoda, M., 62, 64
Jörgensen, D.L., 249

Kasperiuniene, J., 89
Kaulmann, A., 32–33
Kelle, U., 28, 188–189, 291
Kelly, K., 83
Kendall, L., 250
Kendall, P.L., 231–232
Khazal, J., 13
Kirk, J.L., 308
Kitchin, R., 215
Kluge, S., 291
Knoblauch, H., 209
Kozinets, R.V., 251
Krippendorf, K., 265
Kromrey, H., 12, 138–139
Kuhn, T.S., 27
Kuula-Luumi, A., 202
Kvale, S., 245

Lakatos, I., 26
Largan, C., 202
Laswell, H.D., 263
Lazarsfeld, P.F., 62, 182
Lechner, B., 209
Leeman, J., 342–343
Legewie, H., 309
Liamputtong, P., 42
Lillis, G., 341–342
Lincoln, Y.S., 310–312, 313, 336
Lofland, J., 69, 70, 332, 335–336

SUBJECT INDEX

NOTE: page numbers in *italic* type refer to figures. Page numbers in **bold** type refer to tables.